Rubella Viruses

PERSPECTIVES IN MEDICAL VIROLOGY

Volume 15

Series Editors

A.J. Zuckerman

Royal Free and University College Medical School
University College London
London, UK

I.K. Mushahwar

Abbott Laboratories
Viral Discovery Group
Abbott Park, IL USA

Rubella Viruses

Editors

Jangu Banatvala

King's College London Medical School at Guy's
King's College and St Thomas' Hospitals
London, UK

Catherine Peckham

Centre for Paediatric Epidemiology and Biostatistics
UCL Institute of Child Health
London, UK

ELSEVIER

Amsterdam – Boston – Heidelberg – London – New York – Oxford – Paris
San Diego – San Francisco – Singapore – Sydney – Tokyo

Elsevier
Radarweg 29, PO Box 211, 1000 AE Amsterdam, The Netherlands
The Boulevard, Langford Lane, Kidlington, Oxford OX5 1GB, UK

First edition 2007

Library of Congress Cataloguing-in-Publication Data
A catalog record for this book is available from the Library of Congress

British Library Cataloguing in Publication Data
A catalogue record for this book is available from the British Library

ISBN-13: 978-0-444-50634-4
ISBN-10: 0-444-50634-9
ISSN: 0168-7069

For information on all Elsevier publications
visit our website at books.elsevier.com

Transferred to digital print 2007
Printed and bound by CPI Antony Rowe, Eastbourne

Working together to grow
libraries in developing countries
www.elsevier.com | www.bookaid.org | www.sabre.org

ELSEVIER   BOOK AID   Sabre Foundation
            International

Contents

Preface

Since its introduction in the 1960s, rubella vaccination has substantially changed the epidemiology of rubella and congenital rubella nationally and internationally. Worldwide, hundreds of thousands of congenital rubella births have been averted, but the infection continues to inflict a considerable burden in many countries. Different national and regional approaches to rubella control, the ease with which infection can be imported across national boundaries, and the fact that infection is usually mild and frequently asymptomatic, mean that the elimination of rubella will be no easy task. Success in controlling rubella will depend on a sustained political and economic commitment to vaccination and surveillance over the long term. This book highlights the enormous challenges to be faced, through an up-to-date appraisal of the molecular, clinical, laboratory and epidemiological features of rubella and congenital rubella in the vaccine era.

Jangu Banatvala
London, UK

Catherine S. Peckham
Centre for Paediatric Epidemiology and Biostatistics
UCL Institute of Child Health
London, UK

Historical Introduction

Rubella was originally known by its German name "Roteln" having been described originally by two German physicians. For many years German measles was confused with other infections causing a rash, particularly measles and scarlet fever and even today a diagnosis on clinical grounds alone is notoriously inaccurate (see Chapter 2). German measles was, in due course, recognised as a distinct disease by an International Congress in Medicine in London in 1881 and the infection designated as "rubella" was accepted at about that time (reviewed by Best and Banatvala, 2004).

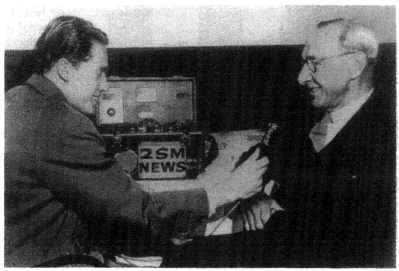

Norman McAlister Gregg being interviewed

It was not until 1941 that an Australian ophthalmologist, Norman McAlister Gregg in his epoch-making paper produced evidence, which showed that if acquired in early pregnancy, rubella caused developmental defects in the fetus (Gregg, 1941). In 1941, there were extensive outbreaks of rubella among troops in training in New South Wales, Victoria and Queensland, being mobilised during the Second World War. Infection spread to the civilian population and it is possible that troops may have infected their wives on returning home prior to serving overseas. Gregg noted the presence of cataracts, usually bilateral, microphthalmia and the "salt and pepper" appearance of a characteristic retinopathy occurring in congenitally infected infants. A high percentage of infants also had cardiac anomalies, failed to thrive and in due course many were found to have severe perceptive deafness.

Gregg was the first to recognise that a virus infection in humans could result in embryopathy and that damage was the result of infection acquired in early fetal life. Although Gregg's findings were not universally accepted initially, other retrospective studies, not only in Australia but in other countries, subsequently confirmed his findings (reviewed by Hanshaw et al., 1985). The earlier studies were retrospective and, as might be expected, were at risk of overestimating the incidence of defects, since the starting point for investigations was the delivery of infants with congenital anomalies, whose mothers reported a rubella-like rash during pregnancy. However, prospective studies carried out in the 1950s and early 1960s, in which the starting point was a mother with a rubella-like illness, rather than an infant with congenital anomalies, revealed a much lower incidence of congenital malformation, varying from 10% to 54% (Hanshaw et al., 1985). But such studies underestimated the incidence of rubella-induced congenital anomalies, since they did not then have the benefit of laboratory confirmation of the clinical diagnosis. Undoubtedly, many women with rash in pregnancy were included and were delivered of healthy babies, but had infections other than rubella. However, subsequent studies showed that following virologically confirmed rubella in the first trimester, the fetus is almost invariably infected, and about 80–85% of infants are damaged (see Chapter 2). Thus, despite the limitations of the earlier retrospective studies, the original observations were a reasonably accurate assessment of the risk to the fetus following maternal rubella, but this was due to the fact that the extensive Australian epidemics in the 1940s were caused by rubella rather than by other infections-inducing rashes.

Rubella virus was isolated for the first time in 1962 by two groups working independently in the USA employing different cell culture techniques (Parkman et al., 1962; Weller and Neva, 1962). This discovery led to the development of serological tests, which could be used to determine rubella virus immunity, and confirm or refute the clinical diagnosis of rubella among women who were exposed to, or who developed rubella during pregnancy. Laboratory investigations were now also available to diagnose, and be used during follow-up studies on infants who had been exposed to rubella *in utero*.

In 1963/1964, the USA experienced one of the most extensive recorded outbreaks of rubella, which led to a greater understanding of the pathogenesis of congenitally acquired rubella as well as a fuller appreciation of the clinical features and sequelae. It was estimated that somewhere of the order of 20,000–30,000 rubella damaged babies resulted from this epidemic and that these infants experienced a generalised and persistent infection, occurring not only *in utero* but extending into infancy (see Chapters 2 and 3). Although it was thought that the wider spectrum of anomalies noted among infants with congenitally acquired rubella following the

1963/1964 epidemic, termed at that time "the expanded rubella syndrome", was a new phenomenon, careful inspection of clinical notes of infants with congenitally acquired disease from previous but smaller outbreaks showed that many of the so-called new clinical features had been observed previously.

The extensive nature of the 1963/1964 rubella epidemic emphasised the importance of attempting to prevent infection by the development of vaccines, and during 1965/1966 attenuated vaccines were developed and the first vaccine trials started.

During the next few years, different vaccines were licensed and rubella vaccination programmes commenced in the USA, UK and subsequently in other parts of the world (see Chapter 4) making congenitally acquired rubella a preventable disease. Current vaccination programmes have reduced the incidence of rubella markedly in many industrialised countries and WHO has now put forward programmes to eliminate rubella, not only in industrialised countries, but also in many developing countries (see Chapters 4 and 5).

Table 1 illustrates some of the principal developments in the history of rubella.

Table 1

Main developments in history of rubella

1881	International Congress on Medicine recognised rubella as a distinct disease
1941	Gregg in Australia recognises teratogenic effects
1962	Rubella virus isolated in cell culture. Neutralisation tests developed
1963–1964	Extensive European and USA epidemics. 12.5 million rubella cases, 11,000 fetal deaths and 20,000 CRS cases in USA
1969 and 1970	Attenuated rubella vaccines licensed in USA and UK (USA universal childhood programme; UK selective vaccination of prepubertal school girls)
1971	MMR licensed in USA
1978, 1979 and 1983	Severe UK rubella epidemics mostly involving adolescent and young adult males but some pregnant women
1986	Rubella virus genome sequenced
1988	UK policy augmented by offering MMR to pre-school children of both sexes
1989	USA introduced a two-dose vaccination at age 12–15 months and at age 4–5 years or 11–12 years
1989–1991	Resurgence of rubella in USA
1996	In UK, schoolgirl vaccination discontinued but second dose of MMR introduced for children aged 4–5 years
2000	WHO recommends immunisation policies for elimination of CRS
2002	123 (57%) of 212 of countries and territories include rubella vaccination in national immunisation programmes

Source: Modified from Lancet 2004; 363: 1127–1137.

References

Best JM, Banatvala JE, Rubella. In: Principles and Practice of Clinical Virology (Zuckerman AJ, Banatvala JE, Pattison JR, Griffiths PD, Schoub BD, editors) Chapter 12, Chichester: Published by John Wiley & sons; 2004; pp. 427–457.

Gregg NM. Congenital cataract following German measles in mother. Trans. Ophthalmol Soc Aust 1941; 3: 35–46.

Hanshaw JB, Dudgeon JA, Marshall WC. Viral Diseases of the Fetus and Newborn. 2nd ed. Philadelphia: W.B. Saunders; 1985.

Parkman PD, Buescher EL, Artenstein MS. Recovery of rubella virus from army recruits. Proc Soc Exp Biol Med 1962; 111: 225–230.

Weller TH, Neva A. Propagation in tissue culture of cytopathic agents from patients with rubella-like illness. Proc Soc Exp Biol Med 1962; 111: 215–225.

Rubella Viruses
Jangu Banatvala and Catherine Peckham (Editors)
© 2007 Elsevier B.V. All rights reserved
DOI 10.1016/S0168-7069(06)15001-6

Chapter 1

Molecular Virology of Rubella Virus

Min-Hsin Chen, Joseph Icenogle
Division of Viral Diseases, Centers for Disease Control and Prevention, 1600 Clifton Road, Atlanta, GA 30333, USA

Introduction

Rubella virus is the sole member of the Rubivirus genus in the *Togaviridae* family, which also includes the Alphavirus genus (Chantler et al., 2001). Alphaviruses and rubella virus share a similar genetic organization and replication strategy. Alphaviruses are transmitted to humans by arthropods while rubella virus is transmitted between humans; humans are the only known hosts for rubella virus. An incomplete list of alphaviruses and the human diseases they are known to cause is Ross River virus (fever, arthritis, rash), Semliki Forest virus (fever, encephalitis), Venezuelan Equine Encephalitis virus (fever, encephalitis), and Sindbis virus (fever, arthritis, rash) (Griffin, 2001). There is little sequence homology between rubella virus and any alphavirus, limiting the possibilities for studying the relationships between rubella virus and alphaviruses by simple phylogenetic analysis. Since rubella virus is more difficult to culture than some alphaviruses (e.g. Sindbis virus), equivalent information for alphaviruses and rubella virus is often lacking. Thus, in the present chapter, some information is presented from work done with alphaviruses and then extrapolations are made to rubella virus.

Rubella virus is a potent, infectious, teratogenic agent, which continues to cause devastating epidemics of congenital rubella syndrome (CRS) in much of the world (Cooper, 1985; Lee and Bowden, 2000). The virus was isolated in 1962 and by 1969 an attenuated, live vaccine was licensed (Parkman et al., 1962; Weller and Neva, 1962). Viruses isolated from CRS cases are identical to viruses from postnatal rubella cases (Lee and Bowden, 2000). *In utero* infection with rubella virus in the first trimester usually produces one or more of a set of specific pathologies in the fetus. Although there have been some molecular studies in cell lines directed toward the pathogenesis of rubella virus, the molecular details of the pathogenesis (i.e. the mechanism(s) of teratogenicity of rubella virus) in humans are poorly understood.

Structure of the virion

Rubella virus virions are particles about 70 nm in diameter with a lipid envelope containing two viral glycoproteins, E1 and E2, and a nucleocapsid, containing the positive-strand RNA molecule and the capsid protein, C (Dominguez et al., 1990; Risco et al., 2003; Zheng et al., 2003c) (Fig. 1A). The structure of the rubella virion has not been precisely determined by fitting crystallographic structures of the individual proteins into cryoelectron microscopic maps, as has been done for alphavirus virions (Zhang et al., 2002; Gibbons et al., 2004). These studies of the virion structures of various alphaviruses have shown interactions between the C protein and RNA, the E1–E2 glycoprotein heterodimers organized into trimers on the virion surface, and the overall structure of the virion. Major conformational changes in the E1–E2 heterodimer occur at low pH and upon interactions with cells, exposing the fusion protein on the E1 protein of Sindbis virus (see Kuhn et al., 2002 for discussion; Paredes et al., 2004). These studies likely provide an approximate model for the structure of the rubella virion and conformational changes that likely occur in the rubella virion (Fig. 2). One should also note that there are significant similarities in structure and function between the large envelope protein of alphaviruses and the envelope protein of flaviviruses, which are more distantly related to the alphaviruses than is rubella virus (Modis et al., 2004).

The C proteins in the rubella virion exist as disulfide-linked homodimers, (although the dimerization is not required for formation of viral particles (Lee et al., 1996)). Analysis of the amino acid sequence of the rubella virus C protein suggests that the N-terminal half of this protein interacts with RNA, because it is hydrophilic and rich in prolines and arginines (reviewed in Frey, 1994). The major RNA-binding domain in the C protein has been located within amino acid residues 28–56, but other regions, including the C-terminus, might also be involved in enhancing the interaction (Liu et al., 1996). The C-terminus of the C protein is very hydrophobic and retains the putative signal peptide of the E2 protein (reviewed in Frey, 1994). Therefore, the C proteins in virions are anchored to the viral membrane by their C-termini, with the N-terminal regions inside the viral envelope and likely contacting the RNA. Since the C proteins are bound to the viral membrane, nucleocapsid assembly/disassembly for rubella virus may well occur by pathways different from those of alphaviruses (see below). It is possible that the N-terminal region of the C protein interacts with the cytoplasmic tail of E1 and is involved in budding (Hobman et al., 1994).

The E1 and E2 glycoproteins exist as heterodimers in the rubella virion. The E1 protein is a class 1 transmembrane protein with three N-linked glycosylation sites in the N-terminal half of the protein. Although glycosylation of E1 does influence the

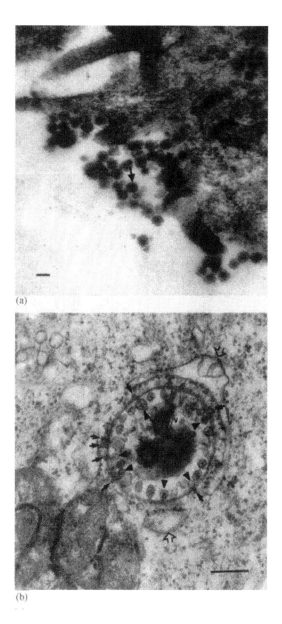

Fig. 1 Electron micrographs from rubella virus infected Vero cell culture. Panel A. Electron micrograph of rubella virus particles in infected Vero cell culture (isolate JV 13R-23) showing dense core and surrounding lipid bilayer (one virion is indicated by arrow). The virus was isolated from human case of rubella. (Electron micrograph by Dr. Elena I. Ryabchikova, Dr. G. I. Tiunnikov, and Dr. Vladimir S. Petrov (State Research Center of Virology and Biotechnology VECTOR, Russia) with permission). Bar corresponds to 60 nm. Panel B. Rubella virus replication complex in Vero cells. Each complex is a modified lysosome-containing vesicles (arrow heads) that line the inner membrane of a cytopathic vacuole (v). The rough endoplasmic reticulum is indicated by open arrows. The bar represents 200 nm. Figure is from Lee et al. (1999), with permission. The closed arrows with tails indicate rubella virus core particles, which were seen in association with the periphery of cytopathic vacuoles in this study, but this observation was not found in another study (Risco et al., 2003).

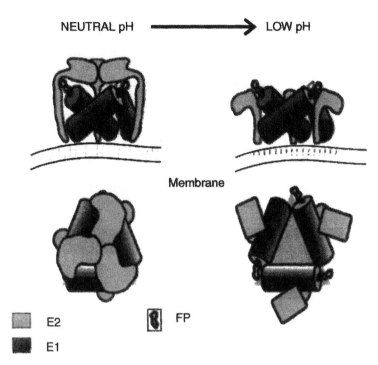

NEUTRAL pH ⟶ LOW pH

Membrane

E2

E1

FP

Fig. 2 Proposed configuration of rubella E1 and E2 proteins on the surface of rubella virus virions by analogy with the alphaviruses. The configuration of the E1 and E2 proteins on the surface of alphavirus virions and the position of the fusion peptide (FP) on the E1 protein at neutral pH, and the proposed configuration at acid pH are shown (e.g. dengue virus). Although no high-resolution structures have been determined for rubella virus virions, since rubella virus virions have a similar composition to that of alphaviruses, it is likely that rubella E1 and E2 proteins will have a similar structure and will undergo similar transitions to those shown. Figure adapted from Kuhn et al. (2002), with permission. Proposed configurations of rubella proteins are the opinion of the authors of this chapter.

formation of infectious virus, likely because glycosylation is required for proper folding of the E1 protein, glycosylation does not play a role in the antigenicity of the virion (Qiu et al., 1992; Ramanujam et al., 2001). The E1 protein contains important functional domains. The fusion peptide, which is exposed during fusion of the virion membrane with cellular membranes during entry into cells, is likely located at amino acid 81–109 of the E1 protein (Qui et al., 2000). Amino acid residues 81–109 of E1 are also likely important for the interaction between E1 and E2 (Yang et al., 1998). The E1 protein is important in the antigenicity of the rubella virus virion; it contains antigenic sites, as defined by monoclonal antibody binding, in the region between amino acids 245 and 284. In patients who have had rubella, a hemagglutination inhibition and neutralization epitope maps to amino acids 208–239 (Ho-Terry et al., 1984; Chaye et al., 1992; Wolinsky et al., 1993; Cordoba et al., 2000). The E2 protein is also a class 1 transmembrane protein, which is heavily glycosylated, both N- and O-linked (Nakhasi et al., 2001). The E2 protein remains linked to the signal peptide of the E1 protein (Hobman et al., 1997). Although the E2 protein of Sindbis is known to be involved in receptor binding, receptor binding has not been mapped to the rubella virus E2 protein (Myles et al., 2003; discussed in Paredes et al., 2004).

There is significant sequence variation among the currently circulating rubella viruses; although a specific antigenic site in the E2 protein recognized a monoclonal antibody in western blots is not conserved, overall there is only one serotype of rubella viruses (Zheng et al., 2003c). There is evidence that the E1 protein alone may be useful as a vaccine (Perrenoud et al., 2004). All evidence indicates that protection from all circulating viruses is achieved by vaccination using the current, most common, live-attenuated vaccine strain of rubella virus, RA27/3.

Genome organization

The rubella virus genome is a single-stranded RNA of 9762 nucleotides (nts) in length (Dominguez et al., 1990), which is 5' capped and 3' polyadenylated and serves as a messenger RNA (mRNA) during infection. One of the significant characteristics of the rubella virus genome is its high $G + C$ content (about 70%), which is the highest of any RNA virus thus far sequenced. The genome contains two open reading frames (ORFs): the 5' proximal ORF encoding nonstructural proteins (NSPs), including a RNA-dependent RNA polymerase that is essential for viral genome replication, and the 3' proximal ORF encoding the viral structural proteins (SP) (the capsid protein and the two envelope glycoproteins) (Fig. 3A). The untranslated regions (UTRs) in the rubella virus genome include a 40-nt sequence at the 5' end (5' UTR), a ~118-nt sequence in the intragenic region between the ORFs and a 59-nt sequence at the 3' end (3' UTR) (reviewed by Frey, 1994).

Attachment and entry

The effects of rubella virus replication on the host cell are dependent on the type of cell infected (Risco et al., 2003). Thus, one must be careful not to extrapolate from one cell line to another and, more importantly, from replication of rubella virus in cell lines to replication of rubella virus in humans. Nevertheless, enough is known about the attachment, entry, and replication of rubella virus and related viruses in cell lines to provide a reasonable overall understanding of rubella virus attachment, entry, and replication.

Attachment of rubella virus to cellular receptors presumably occurs via binding sites on the E2 and/or E1 glycoprotein molecules, even though the cellular receptor for rubella virus has not been identified nor has the receptor-binding site on one or both of the glycoproteins. Both cellular receptors and receptor-binding sites on the glycoproteins have been identified for various alphaviruses (discussed in Kuhn et al., 2002; Myles et al., 2003). Drugs that inhibit specific endocytic pathways have

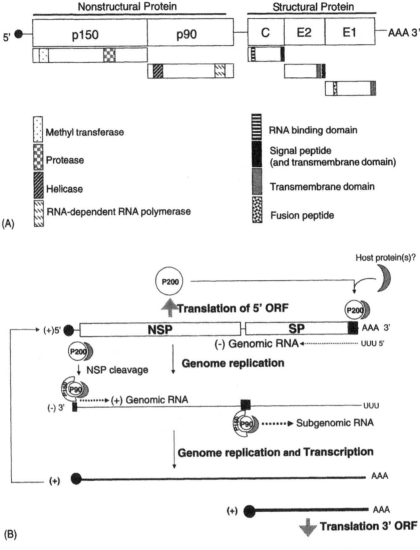

Fig. 3 Rubella Virus Genome Organization (A) and RNA replication (B). (A) The two open reading frames (ORFs) are denoted by open boxes and the three untranslated regions (UTRs) in the genome are indicated by solid lines. The 5' cap structure is indicated by a solid circle and 3' polyadenylation is indicated by AAA. Individual proteins, after processing of the polyproteins from the 5' and 3' ORFs are shown below the genome. Known functional domains in each protein are denoted by boxes. (B) Rubella genome replication. The minus- and plus-strand RNAs are indicated by (−) and (+), respectively. The putative promoters for each RNA species on the plus- or minus-strand genomic RNA are denoted by solid boxes. Translation of 5' and 3' ORFs is indicated by the gray arrow and the progress of RNA synthesis by rubella replication proteins is indicated by arrows with dash lines.

been used to investigate rubella virus entry into Vero cells with the result that the major pathway utilized by rubella virus has been identified as clathrin-mediated endocytosis (also see Lee and Bowden, 2000 and Kee et al., 2004). It is reasonable to assert that a conformational change(s) in the E1 and/or E2 proteins occurs following virus attachment which exposes the fusion peptide in E1, allows the fusion of viral and cellular membrane, and allows the entry of the RNA (or the nucleocapsid) into the cytoplasm (Fig. 2). Since the nucleocapsid is anchored to the viral membrane via the C-termini of its C proteins, it is not clear whether membrane fusion and nucleocapsid disassembly are separate events. In the case of alphaviruses these conformational changes are well understood and occur at low pH (see Kuhn et al., 2002 for discussion).

Intracellular site of RNA replication

Early in infection, rubella virus infected cells contain cytopathic vacuoles which themselves contain RNA-replication complexes (Magliano et al., 1998) (Fig. 1B). The number and size of rubella virus-induced cytopathic vacuoles depend on the type of cell infected (Risco et al., 2003). Reports of colocalization of the C protein with cytopathic vacuoles during RNA replication and the enhancement of poorly replicating, rubella virus self-replicating replicons by the C protein suggest involvement of the C protein in RNA replication (Lee et al., 1994; Chen and Icenogle, 2004). Self-replicating replicons are modified rubella virus RNAs which have a partial genome; usually replicons have the SP ORF replaced by another gene.

Transcription and translation

The genomic RNA of rubella virus can be an mRNA for translation of the NSP ORF, a template for the minus-strand genomic RNA synthesis, or encapsidated by viral SP to form nucleocapsids during assembly. Only the NSP ORF is translated from genomic RNA; a polyprotein of molecular weight ~200 kDa is produced. The NSP ORF encodes the major viral proteins responsible for viral genome replication (i.e. synthesizing plus- and minus-strand genomic RNA) and synthesizing RNAs for translation (viral plus-strand genomic RNA and a subgenomic RNA). Production of viral proteins is dependent on the cellular translational machinery and *cis*-acting elements in the mRNAs are involved in regulation of translation efficiency. The 5′ stem-loop structure, for example, is important for the translation efficiency of the NSP ORF (see Pogue et al., 1993; Pugachev and Frey, 1998a). The effect of a 3′ terminal stem-loop on the translation efficiency is unclear. Enhanced translation of a monocistronic minireplicon by the 3′ stem-loop was observed (Pogue et al., 1993), but no significant effect on viral NSP translation or from subgenomic RNA was seen when using a self-replicating rubella virus replicon (see below; Chen et al., 2004; Pappas et al., 2006). The rubella virus subgenomic

RNA encodes the 3′ proximal ORF (the SP ORF). The subgenomic RNA only serves as an mRNA for the production of viral SP. The 5′ UTR (~76 nts in F-therien) of the subgenomic RNA was found to be necessary for optimal translation efficiency in replicons (Pappas et al., 2006). Although a second in-frame start codon with a consensus Kozak context exists in the SP ORF (located at the 9th amino acid downstream from the start codon of the first ORF), translation of SP ORF is preferentially initiated from the first AUG (authors' unpublished data). In Sindbis virus, a hairpin structure located downstream of the AUG start codon which initiates translation for the C protein, was important for enhancing translation of subgenomic RNA (Frolov and Schlesinger, 1996). A similar hairpin structure in rubella virus C protein coding region was found. Unlike Sindbis virus, this structure attenuates translation from SG RNA (Pappas et al., 2006).

Both genomic and subgenomic RNAs are transcribed by viral replication proteins using the minus-strand genomic RNA as a template. In infected cells, the plus-strand RNA species are present in higher molar ratio than the minus-strand RNA, and more subgenomic RNA is synthesized than plus-strand genomic RNA (molar ratio of 1.6:1) (Hemphill et al., 1988). Interestingly, a study of the 3′ *cis*-acting elements using a self-replicating rubella replicon encoding an antibiotic resistant gene showed that the amount of plus-strand RNA correlated better with the number of cells surviving antibiotics than did the amount of minus-strand RNA, suggesting the synthesis of plus-strand RNA, or transcription, may be the key step of a successful viral replication (Chen et al., 2004).

Genome replication

Replication of the rubella virus genome begins with the synthesis of the NSP polyprotein. This polyprotein is presumed to recognize a promoter at the 3′ terminus of the genome and synthesize a minus-strand genomic RNA, which is proposed to form a double-stranded replicative intermediate with the genomic RNA. The minus-strand genomic RNA is then used as a template for the synthesis of both the genomic RNA and the subgenomic RNA. The subgenomic RNA is synthesized by utilizing an internal promoter (subgenomic promoter) and is colinear with the 3′ terminal 3326 nts of the genomic RNA (reviewed by Frey, 1994) (Fig. 3B).

Regulatory elements in RNA

Comparison of nucleotide sequences of alphaviruses revealed four conserved regions which might be involved in virus replication by functioning as *cis*-acting regulatory signals (reviewed by Strauss and Strauss, 1986). These include the 5′ terminus, a 51-nt domain near the 5′ end, the 21-nt sequence found in the "junction region" upstream from the subgenomic RNA start site and the 19-nt sequence adjacent to the 3′ poly(A) tract. Stretches of nucleotides sharing homology with the first three of these four conserved regions are present in the rubella virus genome

(Dominguez et al., 1990), which suggests the importance of these regions in Togavirus genome replication. Development of a rubella infectious cDNA clone (Wang et al., 1994; Pugachev et al., 1997) and self-replicating replicons containing reporter genes (Tzeng and Frey, 2003; Chen et al., 2004) allows the study of the biological activities of these regulatory elements.

Similar to Sindbis virus, the 5' end of the genomic RNA of rubella virus contains a 14-nucleotide single-stranded leader followed by a stem-and-loop structure (nt 15–65) with a potential pseudoknot structure (reviewed by Frey, 1994). A mutagenic study of this element using a rubella virus infectious clone showed that only the AA dinucleotide at nucleotide 2 and 3 was essential for the viability of the virus; the stem-loop structure was important for the level of production of the NSP proteins but maintenance of the complementary sequences of this structure was not required for plus-strand RNA synthesis (Pugachev and Frey, 1998a).

While mutagenesis in most of the 5' UTR was tolerable, most of the 3' UTR was required for viral viability except the 3' terminal 5 nts and poly(A) tail (Chen and Frey, 1999; Chen et al., 2004). The RNA elements in the E1 coding region were required for efficient replication. For example, the 3' terminal 305 nts were necessary for optimal replication. Maintenance of a GC-rich stem-loop structure at the C-terminus of E1 coding region, which was previously shown to interact with three host factors, was not necessary for virus viability.

The RNA complementary to the ~118-nt intragenic region between 5' and 3' ORFs is presumed to function as a subgenomic RNA promoter. Using a modified rubella virus infectious RNA (Pugachev et al., 2000) and a replicon (Tzeng et al., 2001), the minimal subgenomic RNA promoter was mapped from nt −26 through the subgenomic RNA start site and appears to extend to at least nt + 6 (transcription of subgenomic RNA starts at nt + 1), although a larger region is required for the generation of virus with a wild-type phenotype. Interestingly, while the positioning of the rubella virus subgenomic promoter immediately adjacent to the subgenomic RNA start site is similar to that of alphavirus, it does not include the 21-nt sequence homologous with the alphavirus subgenomic RNA promoter which is located between nt −48 and −23 in the rubella virus genome (Tzeng and Frey, 2002).

Regulatory elements of RNA replication—the roles of nonstructural and structural proteins

The NSP polyprotein, p200, is processed into p150 and p90 by a virally encoded papain-like thiol protease (Marr et al., 1994) (Fig. 3A). The order of these polypeptides is NH2–p150-p90–COOH (Forng and Frey, 1995). Both methyl transferase and protease domains are embedded in p150 while the p90 includes the helicase and RNA-dependent RNA polymerase (reviewed by Frey, 1994). Both Cys-1151 and His-1272 are necessary for the catalytic activity of rubella protease in p150 (Marr et al., 1994; Chen et al., 1996) and the cleavage of p200 between Gly-1300 and Gly-1301 occurs both in *trans* and *cis* (Chen et al., 1996; Liang et al., 2000).

The processing of p200 is crucial for RNA replication (Liang and Gillam, 2000). Mutations abolishing cleavage of p200 resulted in accumulation of minus-strand RNA but no plus-strand RNA synthesis. This indicated that uncleaved p200 could function in minus-strand RNA synthesis, whereas the cleavage products p150 and p90 are required for efficient plusstrand RNA synthesis and suggested that different replication proteins may recognize different promoters. A similar mechanism has been suggested for Sindbis virus (Lemm et al., 1994). The synthesis of minus-strand RNA by p200 is *cis*-preferential (Liang and Gillam, 2001), although rescuing replication-defective mutants by *trans*-complementation of the p200 cleavage products can occur (Wang et al., 2002).

The p90 protein contains a conserved putative RNA-dependent RNA polymerase domain (Kamer and Argos, 1984). There is no direct evidence confirming whether rubella p90 functions as RNA-dependent RNA polymerase although mutation of the conserved GDD tripeptide, which is characteristic of RNA-dependent RNA polymerases, in p90 affected virus viability (Wang and Gillam, 2001). Bacterially expressed rubella recombinant protein containing the C-terminus of p150 and N-terminus of p90 (amino acid residues 1225 and 1664 of p200) was proven to have an RNA-dependent nucleoside triphosphatase activity, suggesting that the N-terminus of p90 could function as an RNA helicase required for unwinding RNA structures during replication (Gros and Wengler, 1996). A retinoblastoma protein-binding motif LXCXE, was also found in, which is also required for efficient virus replication (Forng and Atreya, 1999).

Interestingly, in addition to NSP, the SP of rubella virus was also found to be involved in RNA replication (Tzeng and Frey, 2003; Chen and Icenogle, 2004). The minimal domain in SP required to affect replicon replication was mapped to the C protein. The enhanced genome replication may involve both *trans*- and *cis*-acting elements since either including the C encoding sequences in rubella virus replicons or expressing C proteins in *trans* enhances replicon replication. In one study, it was suggested that the C proteins specifically complemented rubella virus NSP, p150 (Tzeng and Frey, 2003). In another study, replication of mutants with deleterious deletions in the 3′ UTR was enhanced by co-expression of C protein, suggesting that C protein may enhance the RNA replication at the early stage in genome replication (Chen and Icenogle, 2004). Less subgenomic RNA was produced by viruses with mutations in the proposed RNA-binding site of the C protein and resulted in poor interaction with a mitochondria matrix protein p32, suggesting that C protein is also important in regulating the amount of subgenomic RNA synthesis (Beatch et al., 2005).

Interactions between rubella virus proteins and cellular proteins involved in cell cycle regulation

Rubella virus induced cytoplasmic effect and subsequent cell death have been associated with capase-3-dependent programmed cell death and the NSP were associated with rubella virus induced cell death (Pugachev et al., 1997; Pugachev and Frey, 1998b). In addition, p90 has been shown to interact with both the retinoblastoma tumor suppressor protein and the cytokinesis regulatory protein, citron-K kinase (Atreya et al., 1998, 2004a). There is evidence that a balance between cellular survival signals and normal proliferative signals is needed for optimal virus growth (Cooray et al., 2005). Although one must note that molecular evidence for the mechanism(s) of teratogenesis are lacking in humans or animal models, these works in cell lines suggest that interactions between rubella proteins (e.g. p90) and with cellular proteins which regulate cell growth may be the reason why rubella virus is a potent human teratogen (reviewed by Atreya et al., 2004b).

Rubella virus establishes a persistent infection in a number of cell types (Chantler et al., 2001). Given the fact that long-term infection is an important feature of the primary manifestation of rubella (CRS), the molecular mechanism(s) for establishment of persistence deserves more investigation.

Assembly of virions

Alphaviruses such as Sindbis undergo a rather straightforward assembly process in which the glycoproteins E1 and E2 are processed through the Golgi while the nucleocapsids are formed in the cytoplasm, and then nucleocapsids and glycoproteins interact through a budding process at the plasma membrane (Frey and Wolinsky, 1999; Schlesinger, 1999). Rubella virus assembly does not seem to have separate pathways for glycoproteins and nucleocapsids, possibly because the C protein is membrane bound via the E2 signal sequence (Suomalainen et al., 1990).

The three SP of rubella virus are translated from the subgenomic RNA into a polyprotein, which is cleaved into the individual proteins by signalase (Frey and Wolinsky, 1999). The E1 and E2 proteins undergo straightforward glycosylation processes in which the site of *N*-glycosylation is the lumen of the endoplasmic reticulum. The site of *O*-glycosylation of E2 has not been determined, but the E2 protein contains sequences for localization in the Golgi; association of E1 with E2 is required for E1 to leave the endoplasmic reticulum (Yang et al., 1998). The C protein must associate with viral RNA at some point during virion assembly. Assembly (and disassembly) of viral nucleoprotein and thus viral replication is regulated by phosphorylation and dephosphorylation of the C protein; phosphorylation inhibits nucleocapsid formation (Law et al., 2003). A 29-nt segment near the 5′ end of the rubella virus genomic RNA is essential for capsid binding to the RNA (Liu et al., 1996). Some viral assembly occurs at the plasma membrane, but most assembly occurs at intracellular membranes (Chantler et al., 2001). Assembly of rubella virus virions occurs in the Golgi and maturation of virions involving condensation of the nucleocapsids has been observed in the Golgi (Risco et al., 2003). Furthermore, Golgi stacks were also detected near cytopathic vacuoles containing RNA-replication complexes, suggesting an integrated RNA replication and assembly process; as mentioned above, the C protein is known to influence RNA replication (Law et al., 2003; Chen and Icenogle, 2004). Rubella virus virions are released from the infected cell in a cell-type-dependent manner; more infectious particles are released from Vero cells than from BHK cells (discussed in Risco et al., 2003). Since glycosylation, transport, and assembly of rubella proteins all occur in the Golgi, it is difficult to sort out factors essential for each process (discussed in Yao and Gillam, 2000).

Molecular epidemiology of rubella

Despite the fact that rubella virus has a similar genomic organization to the alphaviruses, the phylogenetic relationship between the Rubivirus genus and the Alphavirus genus are distant and the evolutionary events connecting rubella virus with alphaviruses are apparently complicated (Gorbalenya et al., 1991; Koonin et al., 1992; Weaver et al., 1993; Frey and Wolinsky, 1999). However, a number of studies have described currently (or recently) circulating rubella viruses in the world and these viruses fall into a relatively small number of phylogenetic groups (Katow et al., 1997; Zheng et al., 2003a). These rubella viruses group into two virus clades (formerly called genotypes), which differ in nt sequence by about 8–10% (Katow et al., 1997; Frey et al., 1998). Each clade consists of a number of genotypes (Frey et al., 1998; Zheng et al., 2003a). A systematic nomenclature proposed at a WHO meeting in September 2004 consists of two clades (1 and 2) and seven genotypes (designated by clade and a letter designation, e.g. 1C) and three provisional genotypes (1a, 1 g, and 2c) (Fig. 4) (WHO, 2005).

Molecular epidemiology has supported control and elimination of vaccine preventable diseases, e.g. polio and measles. For example, the discovery that vaccine-derived poliovirus recombinants can circulate and cause poliomyelitis contributed to understanding a poliomyelitis outbreak in Hispaniola that started in 2000 (Kew et al., 2002). As another example, in 2001, an outbreak of measles in Venezuela and Columbia was caused by an imported measles virus genotype D9 rather than D6, which had caused previous outbreaks in the region (CDC, 2003a,b).

There is sufficient variability in currently circulating rubella viruses to allow the molecular epidemiology of rubella viruses to support rubella control and elimination activities. For example, clade 2 viruses have not been found circulating in the Americas, and thus clade 2 viruses found in this region are considered importations (WHO, 2000). Some rubella virus genotypes are geographically restricted, allowing sporadic cases caused by these genotypes in other locations to be considered importations (CDC, 2005a). A shift in genotype from 1a and closely related viruses to a mixture of other genotypes (1B–E) occurred in the United States around 1981 (authors' unpublished data). The genotypes in the United States after 1981 are consistent with rubella viruses circulating in neighboring countries between 1981 and 2001. This consistency is even clearer for viruses isolated between 1996 and 2000. This molecular evidence was used to support the conclusion that indigenous rubella and CRS have been eliminated from the United States (Reef et al., 2002; CDC, 2005b; Icenogle et al., in press).

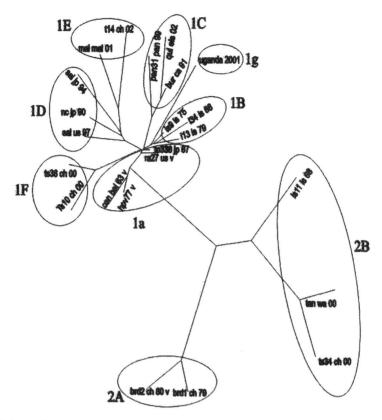

Fig. 4 Phylogenetic tree comparing rubella viruses. A phylogenetic tree using the nucleotide sequence of the entire structural protein-coding region, nucleotides 6481–9741, for viruses representing the indicated genotypes is shown. Rubella viruses are currently classified in 2 clades and 10 genotypes designated 1a, 1B, 1C, 1D, 1E, 1F, 1g, 2A, 2B, and 2c; genotypes which are still provisional are indicated by lower case letters. Nucleotide sequences from the entire structural protein region are not yet available for provisional genotype 2c. The tree shown was created using the Bayesian inference program, MRBAYES. The settings used for the tree were 100,000 ngens; samplefreq, 100; nchains, 4; burnin, 10. Clade credibility values for the tree are high (80 or above) for all nodes except the ra27 us v (RA27/3 vaccine virus) node.

Although the phylogeny and molecular epidemiology of rubella virus is becoming clear, the best operational taxonomic unit(s) for the molecular epidemiology of rubella viruses remains to be determined for some situations. For example, consider rubella virus molecular epidemiology in populations with low vaccine coverage. Multiple genotypes of rubella virus can co-circulate in such populations (Donadio et al., 2003; Zheng et al., 2003b). Rubella viruses in such populations might be very heterogeneous, reflecting many epidemics, importations, etc.; different virus lineages, *within a genotype*, may be useful to track. When tracking viruses within a genotype, extensive sequence information (> 3000 nts) might be needed to fully understand the rubella epidemiology, as has been the case for poliovirus (Liu et al., 2003).

General comments about the molecular virology of rubella virus

Rubella virus is the most important human pathogen in the Togavirus family, still causing at least 100,000 CRS cases in the world each year (Cutts and Vynnycky, 1999; WHO, 2000). It is a very potent infectious teratogenic agent. Rubella virus was intensely studied during the development of rubella vaccines and rubella control in developed countries. Indeed, rubella control and elimination from the United States was a monumental and rapid achievement (CDC, 2005b). Now however, the alphaviruses are more heavily studied largely because these viruses are simply easier to grow in quantity. It is unfortunate, given modern techniques, that more current attention is not given to the molecular virology of rubella virus, an important human pathogen, which continues to circulate uncontrolled in much of the world. There are some intriguing molecular virology problems yet to be solved that are specific to the Rubivirus genus (e.g. molecular pathogenesis and assembly/disassembly).

References

Atreya CD, Kuldarni S, Mohan KVK. Rubella virus P90 associates with the cytokinesis regulatory protein Citron-K kinase and the viral infection and constitutive expression of P90 protein both induce cell cycle arrest following S phase in cell culture. Arch Virol 2004a; 179: 779.

Atreya CD, Lee NS, Forng R-Y, et al. The rubella virus putative replicase interacts with the retinoblastoma tumor suppressor protein. Virus Genes 1998; 16: 177.

Atreya CD, Mohan KVK, Kulkarni S. Rubella virus and birth defects: molecular insights into the viral teratogenesis at the cellular level. Birth Defects Res A Clin Mol Teratol 2004b; 70: 431.

Beatch MD, Everitt JC, Law LJ, Hobman TC. Interactions between rubella virus capsid and host protein p32 are important for virus replication. J Virol 2005; 79: 10807.

CDC (Centers for Disease Control and Prevention). Public health dispatch: absence of transmission of the d9 Measles virus—region of the Americas, November 2002–March 2003. Morb Mortal Wkly Rep 2003a; 52: 228.

CDC (Centers for Disease Control and Prevention). Progress toward measles elimination—region of the Americas, 2002–2003. Morb Mortal Wkly Rep 2003b; 53: 304.

CDC (Centers for Disease Control and Prevention). Global measles and rubella laboratory network, January 2004–June2005. Morb Mortal Wkly Rep 2005a; 54: 1100.

CDC (Centers for Disease Control and Prevention). Achievements in public health: elimination of rubella and congenital rubella syndrome—United States 1969–2004. Morb Mortal Wkly Rep 2005b; 54: 279.

Chantler J, Wolinsky JS, Tingle A. Rubella virus. In: Fields Virology (Fields BN, Knipe DM, Howley PM, editors). Philadelphia: Lippincott: Williams and Wilkins; 2001; p. 963.

Chaye H, Chong P, Tripet B, et al. Localization of the virus neutralizing and hemagglutinin epitopes of E1 glycoprotein of rubella virus. Virology 1992; 189: 483.

Chen J-P, Strauss JH, Strauss EG, Frey TK. Characterization of the rubella virus non-structural protease domain and its cleavage site. J Virol 1996; 70: 4707.

Chen M-H, Frey TK. Mutagenic analysis of the 3′ *cis*-acting elements of the rubella virus genome. J Virol 1999; 73: 3386.

Chen M-H, Frolov I, Icenogle JP, Frey TK. Analysis of the 3′ *cis*-acting elements of rubella virus by using replicons expressing a puromycin resistance gene. J Virol 2004; 78: 2553.

Chen M-H, Icenogle JP. Rubella virus capsid protein modulates viral genome replication and virus infectivity. J Virol 2004; 78: 4314.

Cooper LZ. The history and medical consequences of rubella. Rev Infect Dis 1985; 7(Suppl 1): S2.

Cooray S, Jin L, Best JM. The involvement of survival signaling pathways in rubella-virus induced apoptosis. Virol J 2005; 2: 1.

Cordoba P, Lanoel A, Grutadauria S, Zapata M. Evaluation of antibodies against a rubella virus neutralizing domain for determination of immune status. Clin Diagn Lab Immunol 2000; 7: 964.

Cutts FT, Vynnycky E. Modelling the incidence of congenital rubella syndrome in developing countries. Int J Epidemiol 1999; 28: 1176.

Dominguez G, Wang C-Y, Frey TK. Sequence of the genome RNA of rubella virus: evidence for genetic rearrangement during togavirus evolution. Virology 1990; 177: 225.

Donadio FF, Siqueira MM, Vyse A, et al. The genomic analysis of rubella virus detected from outbreak and sporadic cases in Rio de Janeiro state, Brazil. J Clin Virol 2003; 27: 205.

Forng RY, Atreya CD. Mutations in the retinoblastoma protein-binding LXCXE motif of rubella virus putative replicase affect virus replication. J Gen Virol 1999; 80: 327.

Forng RY, Frey TK. Identification of the rubella virus nonstructural proteins. Virology 1995; 206: 843.

Frey TK. Molecular biology of rubella virus. Adv Virus Res 1994; 44: 69.

Frey TK, Abernathy ES, Bosma TJ, et al. Molecular analysis of rubella virus epidemiology across three continents, North America, Europe, and Asia, 1961–1997. J Infect Dis 1998; 178: 642.

Frey TK, Wolinsky JS. Rubella virus (*Togaviridae*). In: Encyclopedia of Virology (Granoff A, Webster RG, editors). San Diego: Academic Press; 1999; p. 1592.

Frolov I, Schlesinger S. Translation of Sindbis virus mRNA: analysis of sequences downstream of the initiating AUG codon that enhance translation. J Virol 1996; 70: 1182.

Gibbons DL, Vaney M-C, Roussel A, et al. Conformational change and protein–protein interactions of the fusion protein of Semliki Forest virus. Nature 2004; 427: 320.

Gorbalenya AE, Koonin EV, Lai MM-C. Putative papain-related thiol proteases of positive-strand RNA viruses. Identification of rubi- and aphthovirus proteases and delineation of a novel conserved domain associated with proteases of rubi-, alpha- and coronaviruses. FEBS Lett 1991; 288: 201.

Griffin DE. Aphaviruses. In: Fields Virology (Fields BN, Knipe DM, Howley PM, editors). Philadelphia: Lippincott: Williams and Wilkins; 2001; p. 917.

Gros C, Wengler G. Identification of an RNA-stimulated NTPase in the predicted helicase sequence of the Rubella virus nonstructural polyprotein. Virology 1996; 217: 367.

Hemphill ML, Forng RY, Abernathy ES, Frey TK. Time course of virus-specific macromolecular synthesis during rubella virus infection in Vero cells. Virology 1988; 162: 65.

Hobman TC, Lemon HF, Jewell K. Characterization of an endoplasmic reticulum retention signal in the rubella virus E1 glycoprotein. Virology 1997; 71: 7670.

Hobman TC, Lundstrom ML, Mauracher CA, et al. Assembly of rubella virus structural proteins into virus-like particles in transfected cells. Virology 1994; 202: 574.

Ho-Terry L, Cohen A, Tedder RS. Immunological characterisation of rubella virion polypeptides. J Med Microbiol 1984; 17: 105.

Icenogle JP, Frey TK, Abernathy E, Reef SE, Schnurr D, Stewart JA. Genetic analysis of
rubella viruses found in the United States between 1966 and 2004: evidence that indigenous rubella viruses have been eliminated. Clin Infect Dis 2006; 43(suppl 3).

Kamer G, Argos P. Primary structural comparison of RNA-dependent polymerases from
plant, animal and bacterial viruses. Nucleic Acids Res 1984; 12: 7269.

Katow S, Minahara H, Fukushima M, Yamaguchi Y. Molecular epidemiology of rubella by
nucleotide sequences of the rubella virus E1 gene in three East Asian countries. J Infect
Dis 1997; 176: 602.

Kee S-H, Cho E-J, Song J-W, et al. Effects of endocytosis inhibitory drugs on rubella virus
entry into VeroE6 cells. Microbiol Immunol 2004; 48: 823.

Kew O, Morris-Glasgow V, Landaverde M, et al. Outbreak of poliomyelitis in Hispaniola
associated with circulating type 1 vaccine-derived poliovirus. Science 2002; 296: 356.

Koonin EV, Gorbalenya AE, Purdy MA, et al. Computer-assisted assignment of functional
domains in the nonstructural polyprotein of hepatitis E virus: delineation of an additional group
of positive-strand RNA plant and animal viruses. Proc Natl Acad Sci USA 1992; 89: 8259.

Kuhn RJ, Zhang W, Rossmann MG, et al. Structure of dengue virus: implications for
flavivirus organization, maturation, and fusion. Cell 2002; 108: 717.

Law LM, Everitt JC, Beatch MD, et al. Phosphorylation of rubella virus capsid regulates its
RNA binding activity and virus replication. J Virol 2003; 77: 1764.

Lee J-Y, Bowden DS. Rubella virus replication and links to teratogenicity. Clin Microbiol
Rev 2000; 13: 571.

Lee J-Y, Hwang D, Gillam S. Dimerization of rubella virus capsid protein is not required for
virus particle formation. Virology 1996; 216: 223.

Lee J-Y, Marshall JA, Bowden DS. Characterization of rubella virus replication complexes
using antibodies to double-stranded RNA. Virology 1994; 200: 307.

Lee J-Y, Marshall JA, Bowden DS. Localization of rubella virus core particles in Vero cells.
Virology 1999; 265: 110.

Lemm JA, Rumenapf T, Strauss EG, et al. Polypeptide requirements for assembly of functional Sindbis virus replication complexes: a model for the temporal regulation of minus-
and plus-strand RNA synthesis. EMBO J 1994; 13: 2925.

Liang Y, Gillam S. Mutational analysis of the rubella virus nonstructural polyprotein and its
cleavage products in virus replication and RNA synthesis. J Virol 2000; 74: 5133.

Liang Y, Gillam S. Rubella virus RNA replication is cis-preferential and synthesis of neg-
ative- and positive-strand RNAs is regulated by the processing of nonstructural protein.
Virology 2001; 282: 307.

Liang Y, Yao J, Gillam S. Rubella virus nonstructural protein protease domains involved in
trans- and cis-cleavage activities. J Virol 2000; 74: 5412.

Liu H-M, Zheng D-P, Zhang L-B, et al. Serial recombination during circulation of type 1
wild-vaccine recombinant polioviruses in China. J Virol 2003; 77: 10994.

Liu Z, Yang D, Qui Z, et al. Identification of domains in rubella virus genomic RNA and
capsid protein necessary for specific interaction. J Virol 1996; 70: 2184.

Magliano D, Marshall JA, Bowden DS, et al. Rubella virus replication complexes are virus-
modified lysosomes. Virology 1998; 240: 57.

Marr LD, Wang C-Y, Frey TK. Expression of the rubella virus nonstructural protein ORF
and demonstration of proteolytic processing. Virology 1994; 198: 586.

Modis Y, Ogata S, Clements D, Harrison SC. Structure of the dengue virus envelope protein after membrane fusion. Nature 2004; 427: 313.

Myles KM, Pierro DJ, Olson KE. Deletions in the putative cell receptor-binding domain of Sindbis virus strain MRE16 E2 glycoprotein reduce midgut infectivity in *Aedes aegypti*. J Virol 2003; 77: 8872.

Nakhasi HL, Ramanujam M, Atreya CD, et al. Rubella virus glycoprotein interaction with the endoplasmic reticulum calreticulin and calnexin. Arch Virol 2001; 146: 1.

Pappas CL, Tzeng WP, Frey TK. Evaluation of *cis*-acting elements in the rubella virus subgenomic RNA that play a role in its translation. Arch Virol 2006; 151: 327.

Paredes AM, Ferreira D, Horton M, et al. Conformational changes in Sindbis virions resulting from exposure to low pH and interactions with cells suggest that cell penetration may occur at the cell surface in the absence of membrane fusion. Virology 2004; 324: 373.

Parkman PD, Buescher EL, Artenstein MS. Recovery of rubella virus from army recruits. Proc Soc Exp Biol Med 1962; 111: 225.

Perrenoud G, Messerli F, Thierry A-C, et al. A recombinant rubella virus E1 glycoprotein as a rubella vaccine candidate. Vaccine 2004; 23: 480.

Pogue GP, Cao XQ, Singh NK, Nakhasi HL. 5′ sequences of rubella virus RNA stimulate translation of chimeric RNAs and specifically interact with two host-encoded proteins. J Virol 1993; 67: 7106.

Pugachev KV, Abernathy ES, Frey TK. Improvement of the specific infectivity of the rubella virus (RUB) infectious clone: determinants of cytopathogenicity induced by RUB map to the nonstructural proteins. J Virol 1997; 71: 562.

Pugachev KV, Frey TK. Effects of defined mutations in the 5′ nontranslated region of rubella virus genomic RNA on virus viability and macromolecule synthesis. J Virol 1998a; 72: 641.

Pugachev KV, Frey TK. Rubella virus induces apoptosis in culture cells. Virology 1998b; 250: 359.

Pugachev KV, Tzeng W-P, Frey TK. Development of a rubella virus vaccine expression vector: use of a picornavirus internal ribosome entry site increases stability of expression. J Virol 2000; 74: 10811.

Qiu Z, Tufaro F, Gillam S. The influence of *N*-linked glycosylation on the antigenicity and immunogenicity of rubella virus E1 glycoprotein. Virology 1992; 190: 876.

Qui Z, Yao J, Cao H, Gillam S. Mutations in the E1 hydrophobic domain of rubella virus impair virus infectivity but not virus assembly. J Virol 2000; 74: 6637.

Ramanujam M, Hofmann J, Nakhasi HL, Atreya CD. Effect of site-directed asparagine to isoleucine substitutions at the *N*-linked E1 glycosylation sites on rubella virus viability. Virus Res 2001; 81: 151.

Reef SE, Frey TK, Theall K, et al. The changing epidemiology of rubella in the 1990s: on the verge of elimination and new challenges for control and prevention. JAMA 2002; 287: 466.

Risco C, Carrascosa JL, Frey TK. Structural maturation of rubella virus in the Golgi complex. Virology 2003; 312: 261.

Schlesinger MJ. Sindbis and Semliki Forest viruses (*Togaviridae*). In: Encyclopedia of Virology (Granoff A, Webster RG, editors). San Diego: Academic Press; 1999; p. 1656.

Strauss JH, Strauss EG. Structure and replication of the alphavirus genome. In: The Togaviridae and Flaviviridae (Schlesinger S, Schlesinger MG, editors). New York: Plenum Press; 1986; p. 35.

Suomalainen M, Garoff H, Baron MD. The E2 signal sequence of rubella virus remains part of the capsid protein and confers membrane association *in vitro*. J Virol 1990; 64: 5500.

Tzeng W-P, Chen M-H, Derdeyn CA, Frey TK. Rubella virus DI RNAs and replicons: requirement for nonstructural proteins acting in *cis* for amplification by helper virus. Virology 2001; 289: 63.

Tzeng W-P, Frey TK. Mapping the rubella virus subgenomic promoter. J Virol 2002; 76: 3189.

Tzeng W-P, Frey TK. Complementation of a deletion in the rubella virus p150 nonstructural protein by the viral capsid protein. J Virol 2003; 77: 9502.

Wang C-Y, Dominguez G, Frey TK. Construction of rubella virus genome-length cDNA clones and synthesis of infectious RNA transcripts. J Virol 1994; 68: 3550.

Wang X, Gillam S. Mutations in the GDD motif of rubella virus putative RNA-dependent RNA polymerase affect virus replication. Virology 2001; 285: 322.

Wang X, Liang Y, Gillam S. Rescue of rubella virus replication-defective mutants using vaccinia virus recombinant expressing rubella virus nonstructural proteins. Virus Res 2002; 86: 111.

Weaver SC, Hagenbaugh A, Bellew LA, et al. A comparison of the nucleotide sequences of eastern and western equine encephalomyelitis viruses with those of other alphaviruses and related RNA viruses. Virology 1993; 197: 375.

Weller TH, Neva FA. Propagation in tissue culture of cytopathic agents from patients with rubella-like illness. Proc Soc Exp Biol Med 1962; 111: 215.

WHO (World Health Organization). Preventing congenital rubella syndrome. Wkly Epidemiol Rec 2000; 75: 290.

WHO (World Health Organization). Standardization of the nomenclature for genetic characteristics of wild-type rubella viruses. Wkly Epidemiol Rec 2005; 80: 126.

Wolinsky JS, Sukholutsky E, Moore WT, et al. An antibody- and synthetic peptide-defined rubella virus E1 glycoprotein neutralization domain. J Virol 1993; 67: 961.

Yang D, Hwang D, Qui Z, Gilliam S. Effects of mutations in the rubella virus E1 glycoprotein on E1–E2 interaction and membrane fusion activity. J Virol 1998; 72: 8747.

Yao J, Gillam SA. single-amino-acid substitution of a tyrosine residue in the rubella virus E1 cytoplasmic domain blocks virus release. J Virol 2000; 74: 3029.

Zhang W, Mukhopadhyay S, Pletnev SV, et al. Placement of the structural proteins in Sindbis virus. J Virol 2002; 76: 11645.

Zheng D-P, Frey TK, Icenogle J, et al. Global distribution of rubella virus genotypes. Emerg Infect Dis 2003a; 9: 1523.

Zheng D-P, Zhou YM, Zhao K, et al. Characterization of genotype II Rubella virus strains. Arch Virol 2003c; 148: 1835.

Zheng D-P, Zhu H, Revello MG, et al. Phylogenetic analysis of rubella virus isolated during a period of epidemic transmission in Italy, 1991–1997. J Infect Dis 2003b; 187: 1587–1597.

Rubella Viruses
Jangu Banatvala and Catherine Peckham (Editors)
© 2007 Elsevier B.V. All rights reserved
DOI 10.1016/S0168-7069(06)15002-8

Chapter 2

Clinical Features: Post-Natally Acquired Rubella

Jangu E. Banatvala
*King's College London Medical School at Guy's, King's College and St Thomas'
Hospitals, London, SEI 7EH, UK*

Introduction

Infection is transmitted via the aerosol route. Infected persons may excrete high concentrations of virus (e.g. $\geq 10^6$ TCID 50/0.1 ml) although studies on volunteers excreting a rubella vaccine strain (RA27/3) showed that variations in titre as large as a 1000-fold may occur even over a 4–6-h period (Harcourt et al., 1979). Nevertheless, in contrast with infections such as varicella and measles, close and prolonged contact is usually necessary for rubella to be transmitted to susceptible contacts.

Following attachment to receptors likely to be localised in the buccal mucosa and nasopharynx, virus spreads to and replicates in the lymphoid tissue in the nasopharynx and upper respiratory tract. Systemic infection then follows, involving many organs, including the placenta.

Clinical features

Fig. 1 illustrates the relationship between the clinical and virological features of post-natally acquired rubella. Although the incubation period is usually around 13–20 days, this relates to the interval between exposure and onset of rash.

However, lymphadenopathy, which is usually the first clinical manifestation of infection, may precede the onset of rash for up to a week, and persist for up to 10–14 days after the rash has disappeared. The post-auricular and cervical lymph

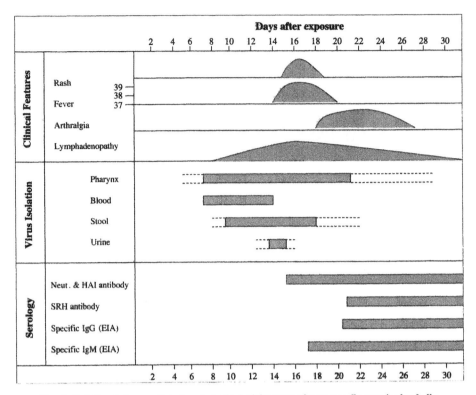

Fig. 1 Relation between clinical and virological features of post-natally acquired rubella.

nodes are usually involved; in adults the lymph nodes are often tender, but less so among children.

In childhood constitutional symptoms are usually mild or absent, and the onset is usually abrupt with the appearance of rash. Adults may develop such prodromal features as fever and malaise, this being associated with viraemia. This is usually of 4–5 days duration and terminates with the development of neutralising antibodies, which develop at about the same time as the rash appears. Viraemia is both lymphocytic and cell free.

The rash, which is at first discreet and presents as pinpoint macular lesions, appears initially on the face and then spreads rapidly to the trunk and limbs. Lesions may coalesce but the rash seldom persists for longer than 3–4 days; in many cases the rash is fleeting. However, rubella may present atypically with an evanescent rash and, in up to 25% of cases, infection may be subclinical.

Since the appearance of rash coincides with the development of humoral immune responses, it has been suggested that this is mediated by antigen–antibody complexes. Rubella virus has been isolated from areas of the skin with rash as well as rash-free areas (Heggie, 1978).

Virus may be recovered from nasopharyngeal secretions for about a week before the onset of rash and for a similar period, but sometimes longer, after the rash has disappeared. Virus may also be detected in the blood, stool and urine, but these are not favoured sites for recovering virus for diagnostic purposes (see Chapter 3).

Differential diagnosis

Since rubella may present atypically or with non-specific symptoms and signs that may be caused by other viruses which do not have a teratogenic potential, it is important that the clinical diagnosis be confirmed by laboratory investigations, particularly during pregnancy. Rubella-like rashes may be induced by such viruses as parvovirus B19, enteroviruses, adenoviruses and such arboviruses as Ross River and Chikungunya. Arthralgia may be a feature of some of these infections, e.g. parvovirus B19, Ross River and Chikungunya. In developing countries, in addition to the above infections, rubella may also be readily confused clinically with dengue and measles, and differentiating these infections is of importance during surveillance programmes involved in rubella elimination (see Chapter 4). Table 1 shows the geographical distribution of some virus infections which may be difficult or impossible to distinguish from rubella.

Complications

Joint involvement is the commonest complication of naturally acquired rubella as well as following rubella vaccination. This usually develops as rash subsides and, although uncommon among males and pre-pubertal females, may occur in up to 50% of post-pubertal females. Symptoms vary in severity, ranging from a transient arthralgia with joint stiffness, to a frank arthritis with pain, limitation of movement and swelling. The finger joints, wrists, knees and ankles are most frequently affected. Generally, symptoms persist for 3–4 days, but may last for up to a month and even occasionally longer, exhibiting a fluctuating course. The pathogenesis is the subject of some debate. Hormonal factors are likely to be involved, since in addition to occurring commonly in post-pubertal females, studies among vaccinees have shown that joint symptoms are likely to develop within 7 days of the onset of the menstrual cycle (Harcourt et al., 1979). Virus has been isolated from the synovial fluid of patients with naturally acquired and vaccine-induced arthritis (Fraser et al., 1983; Best and Banatvala, personal communication) and since symptoms develop when rash subsides and humoral antibodies appear, immune complexes may play a part in pathogenesis. Some investigators have suggested that persistent rubella infection may be associated with chronic arthritis (Chantler et al., 1985; Smith et al., 1987) but a study in which synovial fluid and biopsies of synovial membrane from adults were examined by RT-PCR as well as by virus recovery in cell culture, failed to show an association with chronic inflammatory joint disease (Bosma et al., 1998).

Table 1

Differential diagnosis of postnatal rubella in different geographical regions

Virus infection	Geographical distribution								Key features
	Africa	Asia	Australia	Europe	North America	Central America	South America	Pacific	
Rubella	+	+	+	+	+	+	+	+	
Parvovirus B19	+	+	+	+	+	+	+	+	Erythema infectiosum
Human herpes viruses 6 and 7	+	+	+	+	+	+	+	+	Exanthem subitum, mainly <2 years
Measles	+	+	+	+	+	+	+	+	Prodrome with cough, conjunctivitis, coryza
Enteroviruses	+	+	+	+	+	+	+	+	Echovirus 9, Coxsackie A9 most frequent
Dengue	+	+	+	−	−	+	+	+	Joint and back pain, haemorrhagic complications in children
West Nile fever	+	+	−	+	+	−	−	−	Joint pain
Chickungunya	+	+	−	−	−	−	−	−	Joint pain
Ross River	−	−	+	−	−	−	−	+	Joint pain
Sindbis	+	+	+	+	−	−	−	−	Joint pain

Source: Reproduced from Banatvala and Brown (2004).

In cases in which joint symptoms persist or relapse occurs, parasthesiae, sometimes associated with carpal tunnel syndrome, may develop.

CNS complications are rare. Post-infectious encephalopathy or encephalomyelitis occur in 1:5000 to 1:10,000 cases, symptoms usually developing about a week after the onset of rash (reviewed by Frey, 1997). In general, recovery occurs within 7–30 days, although the course may be prolonged death is unusual. Symptoms are similar to those among patients with other viral encephalitidies, although patients may not be febrile. There is a lymphocytic pleocytosis but intrathecal rubella virus antibody synthesis has not been reported, although rubella virus RNA has been detected in the CSF by reversed transcription PCR (Date et al., 1995). Although death is very rare, post-mortem studies show that, in contrast with other post-infectious encephalitidies, demylination is not present, which suggests that immune mechanisms are not involved in its pathogenesis.

Gullain–Barré syndrome, although an even rarer complication of rubella, has been reported (Atkins and Esmonde, 1991) and consequently it is worth investigating patients lest they have experienced recent infection by rubella, bearing in mind that many adults experience subclinical infections.

Major haematological abnormalities are unusual. Haemolytic anaemia has been reported during outbreaks in Japan (Ueda et al., 1985). As with other exanthemata, a transient depression of thrombocytes may occur, but a purpuric rash is uncommon, occurring in about 1:1500 cases (Steen and Torp, 1956; Ueda et al., 1985). It is possible that thrombocytopenia may result from enhanced clearance of platelets by the reticuloendothelial system, which may be the result of an auto-immune phenomenon (Rand and Wright, 1998).

Unlike measles and varicella, those with impaired cell-mediated immunity, have not been reported to suffer unduly severely from rubella. Thus although the American Academy of Pediatrics, while not recommending the use of live-attenuating vaccines to severely immunocompromised HIV-infected children, express the view that MMR vaccine should not be withheld from those with no evidence of immunosuppression or those who have only a moderate amount of immune suppression (American Academy of Pediatrics, 1999).

There has, however, been a report of a child being inadvertently vaccinated in remission from acute lymphoblastic leukaemia who developed chronic arthritis, and persistent viraemia, despite developing high antibody titres (Geiger et al., 1995).

Congenitally acquired infection

Pathogenesis of congenital rubella

If maternal rubella is acquired during the first trimester of pregnancy, particularly before 8 weeks, this is likely to result in a generalised and persistent infection with multi-system disease.

The severity of infection following maternal rubella in the first trimester, reflects the inefficiency of the placental barrier as well as the inability of the fetus to mount an effective immune response to eliminate rubella virus.

The mechanism by which fetal damage occurs is probably multi-factorial, being the result of rubella-induced cellular damage and the effect of the virus on dividing cells. The fetus is at risk during the period of maternal viraemia at which time placental infection is likely to occur. Tôndury and Smith (1966) in their classical histopathological study showed that following maternal rubella in early pregnancy, the presence of focally distributed areas of necrosis in the epithelium of the chorionic villae and in the endothelial cells of its capillaries. These endothelial cells appeared to be desquamated into the lumen of the vessels, suggesting that virus is transported into the fetal circulation in the form of rubella-infected endothelial cell "emboli" which may result in the blockage of small blood vessels and, if transported into the fetal circulation, result in infection and damage occurring in various fetal organs. The most extensive histological changes occur in early pregnancy, i.e. before fetal defence mechanisms are sufficiently mature to be activated. Thus, a major characteristic of rubella embryopathy in the early weeks of gestation is the presence of cell necrosis in the absence of an inflammatory response.

In addition to cellular damage, there is evidence of retardation in cellular replication, thus infected cells have a reduced lifespan (Rawls and Melnick, 1966), indeed it has been shown that the organs of rubella-infected infants are smaller and contain a subnormal number of cells when compared with those of healthy infants (Naeye and Blanc, 1965). Although virus may spread from infected cells to progeny cells during division, resulting in infected clones of cells (Woods et al., 1966; Rawls et al., 1968), only discreet foci can be detected in a rubella-infected conceptus. The number of infected cells is small (about ∼0.1%) and such infected foci are protected from maternal and fetal immune responses.

During the second trimester, fetal immune responses gradually mature and some transfer of maternal rubella-specific IgG occurs, whereas prior to this, placental transfer is inefficient and the fetus is consequently defenceless against infection. Fetal immunoglobulin secreting plasma cells increase progressively during the second trimester and may be detected after about 15 weeks of gestation (reviewed by Webster, 1998).

Cell-mediated immune responses are involved in the elimination of cells infected by such enveloped viruses as rubella. Cytotoxic T cells, natural killer cells and monocytes, together with the secretion of lymphokines, are involved, but such responses like humeral responses do not mature until the second trimester. Although T cells may be detected as early as the 12th week of gestation, adult expression is delayed for another 5–6 weeks (Lobach et al., 1985), and lymphokine production is immature. Thus, during the second trimester and subsequently, the fetus is afforded protection by the progressive development of immune responses, although *in vitro* studies indicate that fetal organs are still susceptible to rubella (Best et al., 1968). However, rubella virus itself can have an adverse effect on immune responses, since a feature of the congenital rubella syndrome (CRS) neonatally, is the presence of a dysgammaglobulinaemia, with either an increased or a decreased level of immunoglobulins.

Infants with congenitally acquired infection may develop manifestations of late-onset disease in later childhood or as adults. The endocrine organs, the eyes, hearing and the CNS may be affected and behavioural problems, including autism, may occur. Autoimmune phenomena may be responsible for endocrine and CNS disturbances. Rubella virus has been recovered from a cataract removed from a 3-year-old child (Menser et al., 1967a), rubella antigen from the thyroid of a 5-year-old child with Hashimoto's disease (Ziring et al., 1977) and virus has been recovered from the brain of a 12-year-old child developing rubella panencephalitis (Cremer et al., 1975; Weil et al., 1975).

Risks of congenital rubella

Following maternal rubella in the first trimester

The outcome of maternal rubella infection in the first trimester is very varied, as Fig. 2 below illustrates. However, following virologically confirmed rubella in the first trimester, the fetus is almost invariably infected, and in about 80–85% of cases affected manifesting a wide spectrum of anomalies (Miller et al., 1982; Wolinsky, 1996).

Fig. 3 correlates the range of congenital anomalies with gestational age, among 376 infants with congenitally acquired rubella following the extensive 1964 rubella epidemic in the USA. Cardiac and eye defects occurred during the critical phase of organogenesis in the first 8 weeks of pregnancy, whereas retinopathy and deafness were more evenly distributed through the first 16 weeks of gestation.

Studies on autopsy material obtained from infants dying in early infancy from severe and generalised infections, show that virus may be recovered from virtually every organ of the body. Neonatally, virus may be recovered from nasopharyngeal

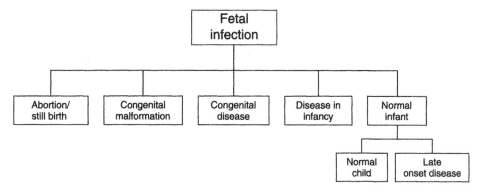

Fig. 2 Possible outcomes of congenital rubella.

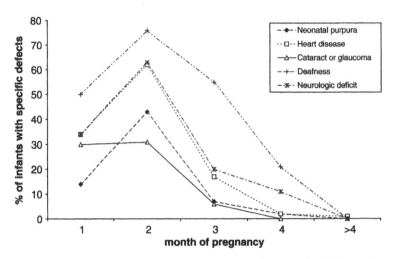

Fig. 3 Relationship between the gestational age and maternal infection and clinical manifestations of clinical rubella. With permission from John Wiley & Son.

secretions, urine, stools and tears. Among infants with congenitally acquired rubella, most are excreting virus at birth but by 3 months, and 9–12 months, the proportion has declined to 50–60% and 10%, respectively. High titres of virus are excreted at birth and such infants may readily transmit infection to susceptible contacts. The mechanism by which rubella persists throughout gestation and for a limited period during the first year of life, remains to be clearly established. Possible mechanisms may involve defective cell-mediated immunity. Thus, Buimovici-Klein et al. (1979) demonstrated diminished lymphoproliferative responses to phytohaemagglutin and rubella antigen and diminished interferon synthesis in children aged 1–12, this being related to the gestational age at which maternal infection occurred. It was greatest during the first 8 weeks. Failure to mount a lymphoproliferative response may be useful diagnostically in determining whether children were infected *in utero* if being investigated after the first year of life, at which time virus excretion and virus-specific IgM responses would no longer be detectable (O'Shea et al., 1992). Alternatively, as stated above, the possibility that a number of virus-infected cells result in infected clones which have a limited life span and when these cells no longer survive, virus excretion terminates (reviewed by Banatvala, 1977; Banatvala and Best, 2000). Selective immune tolerance to the viral E1 protein has also been suggested (Mauracher et al., 1993).

Post-first trimester

In contrast with infants whose mothers acquire rubella in the first trimester, rubella is seldom isolated from infants whose infection is acquired at a later gestational age. Nevertheless, serological studies confirm that about one-third of infants whose mothers were infected between the 16th and 20th week of pregnancy, were infected since rubella-specific IgM is present neonatally (Craddock-Watson et al., 1980*)*. However, as organogenesis is complete by the end of the first trimester, and fetal immune responses, which may limit or terminate infection, mature during the second trimester, infected infants rarely have severe or multiple anomalies.

An analysis of four studies conducted in different countries showed that between 13 and 16, 17 and 20 and after 20 weeks, the proportion with rubella-induced defects were 16.7%, 15.9% and 1.7%, respectively (Banatvala and Best, 2000). In general, deafness, which was usually the sole manifestation of infection, and retinopathy, which *per se* does not usually affect vision, were the only anomalies generally associated with maternal rubella after the first trimester. This stresses the importance of careful assessment of infants with post-first trimester serological evidence of intrauterine infection who appear to be unaffected in order to detect hearing defects as early as possible.

Pre-conception

Although there have been occasional reports on maternal infection occurring as long as 21 days prior to conception published prior to 1984 (reviewed by Best and Banatvala, 2000), a prospective study conducted in Britain and Germany showed that pre-conceptional rubella did not result in transmission of rubella to the fetus. Thus, there was no serological or clinical evidence of intrauterine infection in 38 infants whose mother's rash appeared before or within 11 days after their last menstrual period (LMP), although after an interval of 12 days had elapsed between the LMP and the rash, fetal infection occurred. All of the 10 mothers developing rash 3–6 weeks after their LMP resulted in viral transmission to the fetus (Enders et al., 1988).

Clinical features of congenital rubella

In addition to the spectrum of anomalies, which may be apparent at birth or during the early months of life, congenitally acquired infection may result in a considerably delayed onset of disease manifestations which may not become apparent until adolescence or even later. Thus, congenitally acquired rubella should be regarded as a chronic disease, although it may not always be possible to detect virus in late-onset disease.

The spectrum of manifestations of congenitally acquired rubella is variable according to the gestational period during which maternal rubella is acquired. Congenital disease may only affect, for example, hearing, particularly if acquired post first trimester. If acquired at this stage or even later during gestation, there may be serological evidence of infection but without evidence of disease, although long term follow up is advisable lest late onset manifestations occur.

A wide spectrum of congenital anomalies occurs when maternal rubella is acquired during early gestation, particularly if acquired during the critical phase for organogenesis. Hearing, eye and cardiac defects predominate but other anomalies may also occur. The term congenital rubella syndrome (CRS) is used for infants with such anomalies. Some of these are listed in Table 2 together with the time at which they are commonly recognised. Early and transient features are self-limiting, but permanent and developmental defects which, although taking months or even years before becoming apparent, persist indefinitely. Virtually every organ may be

Table 2

Clinical features of congenital rubella syndrome

	Time when signs commonly recognised[a]	Early transient features	Permanent features[b]
Ocular defects			
Cataracts (unilateral or bilateral)	Early infancy	−	+
Glaucoma	Early infancy	−	+
Pigmentary retinopathy	Early infancy	−	+
Microphthalmia		−	+
Iris hypoplasia		−	+
Cloudy cornea		+	−
Auditory defects			
Sensorineural deafness (unilateral or bilateral)	Early infancy	−	+
Cardiovascular defects			
Persistent ductus arteriosus	Early infancy	−	+
Pulmonary artery stenosis	Early infancy	−	+
Ventricular septal defect	Early infancy	−	+
Myocarditis		+	−
CNS defects			
Microcephaly	Neonatal	−	+
Psychomotor retardation		−	+
Meningoencephalitis	Neonatal	+	−
Behavioural disorders		−	−
Speech disorders			
Intrauterine growth retardation		+	−
Thrombocytopenia, with purpura	Neonatal	+	−
Hepatitis/hepatosplenomegaly	Neonatal	+	−
Bone lesions	Neonatal	+	−
Pneumonitis		+	−
Lymphadenopathy		+	−
Diabetes mellitus		−	+
Thyroid disorders		−	+
Progressive rubella panencephalitis		−	+

Source: Reproduced from Banatvala and Brown (2004).
[a]Time when signs seen commonly used to clinically confirm CRS cases.
[b]Some recognised as late as during adolescence or as adults.

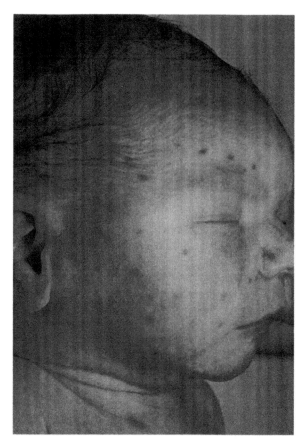

Fig. 4 Purpuric rash in infant with CRS.

affected. Fig. 4 shows the characteristic purpuric rash ("blueberrymuffin" appearance) often apparent at birth in infants whose mothers acquired infection in early pregnancy and who usually have or develop other clinical features such as cardiovascular, eye or hearing anomalies. This rash is associated with a low platelet count and usually spontaneously resolves within a few weeks.

Infants aged between 3 and 12 months may develop a chronic rubella-like rash, pneumonitis and diarrhoea, which are sometimes referred to as "late-onset disease". Although there is a high mortality rate, infants often improve with treatment with corticosteroids and it has been suggested that immune complexes may be responsible for this syndrome.

Cardiovascular anomalies

A patent ductus arteriosus (PDA) and septal defects in association with pulmonary artery and pulmonary valvular stenosis are common and responsible for much of the high perinatal mortality associated with congenital rubella. Fallot's tetralogy may also occur, but more rarely. Some infants also have evidence of neonatal myocarditis (Hastreiter et al., 1967). Rubella may also result in damage and pro-liferation of the intimal lining of the arteries, causing obstructive lesions of the larger- and medium-sized arteries in the systemic and pulmonary circulation. This may result in stenosis of coronary, cerebral and renal arteries (Fortuin et al., 1971; Sever et al., 1985); renal obstruction may result in hypertension (Menser et al., 1967b; Menser and Reye, 1974).

Sensorineural hearing loss

Hearing defects of varying degrees of severity (unilateral and bilateral) are the most common manifestations of congenital rubella, which may be present in 70–90% of cases of children with rubella-associated defects. The burden of deafness due to congenital rubella has almost certainly been underestimated. Congenital rubella is considered to be the most important cause of non-genetic congenitally acquired hearing loss in countries, which are not implementing rubella vaccination pro-grammes (Smith, 1999). Rubella may at first infect the labyrinth, although clinical evidence of vestibular malfunction is uncommon. Rubella then passes through the stria vascularis to infect the sensorial cells of the organ of corti, which results in progressive sensorial hearing loss (Scasso et al., 1998).

Some infants, although having apparently normal hearing in infancy, may subsequently develop sensoneural deafness, which may increase over time. They should therefore be followed up regularly. There has been a report of sudden onset deafness in a child of aged 10 who had apparently normal hearing until then (Sever et al., 1985).

Newer methods are now used to assess hearing in infancy, which includes otoacoustic emissions and auditory brainstem responses (Mehl and Thomson, 1998). Infants delivered of mothers with a history of maternal rubella should therefore be screened by such techniques in early infancy since hearing defects may often be detected much earlier than hitherto in the neonatal period (van Straaten et al., 1996). They will not be detected if it is progressive so caution is needed. Prompt recognition and management is of importance in order to reduce delays in language development and social adaptation. Unfortunately the equipment is costly and its reliability in field settings has yet to be determined. This currently limits its ap-plication in developing countries where congenital rubella defects continue to be a major problem.

Eye defects

The major defects first described in Gregg's epoch-making paper in 1941 (Gregg, 1941), included cataract, which was usually bilateral and filled the pupillary area, and microphthalmia (Fig. 5). Rubella virus may affect not only the lens, but also the retina and ciliary body. Apart from retinopathy, cataracts, which can be

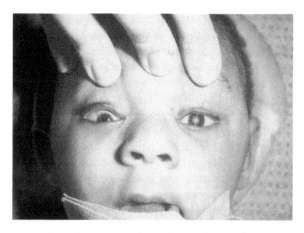

Fig. 5 Congenital rubella cataract in an infant aged 9 months.

unilateral, have been shown to be the most common manifestation of eye defects caused by congenital rubella. However, they are a relatively rare manifestation of congenital infection because the critical time of exposure is small. Retinopathy described as "a salt and pepper appearance in the retina", occurs in the majority of patients but may be obscured by the presence of cataracts; eye defects may be associated with strabissmus and nystagmus. There have been reports of late-onset retinal neovascularisation (Deutman and Grizzard, 1978). Glaucoma, although rare in early infancy, may develop in adolescence (Boger, 1980), and may be associated with cataracts and microphthalmia (Givens et al., 1993). Aphakic glaucoma may also occur following cataract aspiration. Corneal abnormalities, including neonatal transient corneal opacities, corneal hydrops and keratoconus have been described (Boger et al., 1981).

Rubella has been reported to be an important cause of visual defects, including blindness in developing countries, including India (Eckstein et al., 1996), Jamaica (Moriarty, 1988) and Ghana (Lawn et al., 2000).

Central nervous system

Involvement of the brain may result in vascular lesions caused by endothelial damage as described above as well as direct viral involvement of brain cells (Chantler et al., 1995). MRI scanning shows reduced grey matter and enlargement of the lateral ventricles (Lim et al., 1995). Further studies employing investigations by ultrasonography and computerised tomographic examination on a small number of patients showed intracranial calcification and linear hyperechogenicity over basal ganglia. Such abnormalities appeared to be predictive of the development of microcephaly (Chang et al., 1996).

Neonatal manifestations include an encephalitis, some infants exhibiting lethargy or irritability and seizures. Microcephaly may be present at birth or develop subsequently. On follow up, infants affected by congenital rubella infection may develop evidence of varying degrees of psychomotor retardation. Behavioural disorders, including autism and a schizophrenia-like disease may develop subsequently (Cooper, 1985). A higher risk of non-affective psychosis among those exposed to rubella *in utero* during the first trimester, but having normal intelligence, has been described (Brown et al., 2000). This report indicated that these disorders were unrelated to such factors as deafness, ethnicity or selection bias.

Progressive rubella panencephalitis (PRP) is a rare manifestation of late-onset disease and about 50 cases have been described. The outcome is invariably fatal (Frey, 1997). Very rarely, PRP may be a complication of post-natally acquired rubella. Clinically, the presentation and disease progression is similar to measles-induced subacute sclerosing panencephalopathy. Histopathological studies have shown enlargement of the lateral ventricles, loss of glial cells and demyelination. The disease probably has an autoimmune aetiology, perhaps as a result of molecular mimicry between viral and host epitopes (Frey, 1997). Serological studies shown raised antibody titre to rubella in the blood and CSF, with oligoclonal bands of rubella-specific IgG in the CFS. Rubella virus itself is somewhat illusive and may be present in a defective form. However, the virus has been isolated following co-cultivation of patients' brain cells.

Endocrine disorders

Type 1 (insulin-dependent) diabetes mellitus (IDDM) is the most frequent late-onset manifestation of congenital rubella after sensorineural hearing loss, first highlighted during follow-up studies of those affected in the 1939–1941 Australian extensive rubella epidemic. In all, 20% had developed IDDM by the time they were in their 30 s (Menser et al., 1978). Follow-up studies on patients with congenital rubella infection acquired during the 1963–1964 epidemic in the USA reported similar findings, showing that 12.4% had developed IDDM but also showed that 20% had pancreatic islet cell cytotoxic surface antibodies (Ginsberg-Fellner et al., 1984). The presence of these antibodies, together with a significantly increased HLA-DR3 and decreased HLA-DR2 haplotype which are associated with autoimmune disorders, are indicative that such mechanisms are involved in the pathogenesis of associated IDDM with congenital rubella infection (Ginsberg-Fellner et al., 1984).

That autoimmune mechanisms are involved in the pathogenesis of IDDM, is supported by the findings of experimental studies which showed that monoclonal antibodies for rubella virus capsid protein recognise B-cell epitopes on human and rat islet cells (Karounos et al., 1993), and that T-cell clones from patients with congenital rubella elicit cytotoxic responses to GAD65, a B-cell autoantigen (Ou et al., 1999).

There have been no reports of rubella virus having been detected in pancreatic tissue from patients with congenital rubella and this may reflect the shortage of suitable autopsy material for investigations. In any case, the precipitating infection inducing autoimmune phenomena may no longer be present. However, it is possible that viral RNA may persist in greatly reduced genome copies but retaining some biological activity following clearance of viral infectivity. Although it has been postulated that such virus infections as enteroviruses may be involved in the pathogenesis of IDDM (reviewed by Minor and Muir, 2004) evidence is as yet inconclusive and only rubella virus infections, if congenitally acquired, has been shown to be involved in the pathogenesis of IDDM among humans. Consequently, further studies using this model may be of importance in elucidating the mechanisms involved in the pathogenesis of IDDM.

Thyroid disorders, including hypothyroidism, hyperthyroidism and thyroiditis, occur in about 5% of patients with congenital rubella defects. As with IDDM, autoimmune phenomena are likely to be involved, since antibodies to the globulin and microsomal fractures of the thyroid were shown to be present in 23% of the 201 deaf patients with congenital rubella (Clarke et al., 1984). Some patients may have evidence of both IDDM and thyroid disorders (Floret et al., 1980).

Although growth hormone deficiency may be responsible for poor post-natal growth, there have been conflicting reports about the integrity of the hypothalmic/ pituitary axis (Preece et al., 1977; Oberfield et al., 1988). Japanese studies showed that growth retardation correlated with the severity of clinical manifestations at birth being more severe among those exposed to maternal rubella in the first week of gestation, particularly among patients with rubella-associated defects who had cataracts. In contrast, those infected at a later gestational age, although having some evidence of growth retardation in infancy and early childhood, had a normal growth pattern after starting school (Tokugawa et al., 1986).

Long-term prognosis. The most extensive series of patients studied are those with congenitally acquired infection occurring during the Australian epidemic in the 1940s. Most were of normal stature, only 15% being less than the third percentile for height (McIntosh and Menser, 1992). Despite most having multiple defects, all of the 40 examined were deaf; they had adjusted socially extremely well. Twenty nine (58%) were married, and between them had 51 children. Only one child was deaf, but this may well have been hereditary, since his father, who was not affected by congenital rubella, was also deaf. However, by 60 years of age, this cohort had an increased prevalence of type 2 diabetes compared with the general population. Follow-up studies had recorded such other disorders as an early menopause (73%) and osteoporosis (13%), all being significantly higher than found in the general Australian population (Forrest et al., 2002).

These results contrast with follow-up studies on 300 congenital rubella survivors of the extensive 1963–1964 epidemic in the USA. Only one-third were leading a normal life and the remaining needed care, varying from having to live with relatives or requiring 24-h care. Differences between the two studies may, in part, be the result of advances in clinical management following the 1963–1964 outbreak, reducing mortality, particularly from congenital heart disease, which, during the 1940s, was not possible. However, such apparent improvement in prognosis in infancy may have resulted in survivors developing severe long-term sequelae.

References

American Academy of Pediatrics. Committee on infectious diseases and committee on pediatric AIDS. Measles immunization in HIV infected children. Pediatrics 1999; 103: 1057–1060.

Atkins MC, Esmonde TF. Guillain–Barre syndrome associated with rubella. Postgrad Med J 1991; 67(786): 375–376.

Banatvala JE. Health of mother, fetus and neonate following maternal infections during pregnancy. In: Infections and Pregnancy (Coid CR, editor). London: Academic Press; 1977; pp. 437–488.

Banatvala JE, Best JM. Rubella. In: Principles and Practice of Clinical Virology (Zuckerman AJ, Banatvala JE, Pattison JR, editors). 4th edition. Chichester: Wiley; 2000.

Banatvala JE, Brown DWG. The Lancet 2004; 363: 1127–1137.

Best JM, Banatvala JE. Personal Communication.

Best JM, Banatvala JE. Rubella. In: Principles and Practice of Clinical Virology (Zuckerman AJ, Banatvala JE, Pattison JR, editors). 4th edition. Chichester: Wiley; 2000.

Best JM, Banatvala JE, Moore BM. Growth of rubella virus in human embryonic organ cultures. J Hyg (Camb) 1968; 61: 407–413.

Boger III WP. Late ocular complications in congenital rubella syndrome. Ophthalmology 1980; 87: 1244.

Boger III WP, Petersen RA, Robb RM. Keratoconus and acute hydrops in mentally retarded patients with congenital rubella syndrome. Am J Ophthalmol 1981; 91: 231.

Bosma TJ, Etherington J, O' Shea S, et al. Rubella virus and chronic joint disease: is there an association. J Clin Microbiol 1998; 36: 3524–3526.

Brown AS, Cohen P, Greenwald S, Sussex E. Nonaffective psychosis after prenatal exposure to rubella. Am J Psychiatry 2000; 157: 3.

Buimovici-Klein E, Lang PH, Ziring PR, et al. Impaired cell mediated immune response in patients with congenital rubella; correlation with gestational age at time of infection. Pediatrics 1979; 64: 620–626.

Chang YC, Huang CC, Liu CC. Frequency of linear hyperechogenicity over the basal ganglia in young infants with congenital rubella syndrome. Clin Infect Dis 1996; 22(3): 569–571.

Chantler JK, Smyrnis L, Tai G. Selective infection of astrocytes in human glial cell cultures by rubella virus. Lab Invest 1995; 72: 334–340.

Chantler JK, Tingle AJ, Petty RE. Persistent rubella virus infection associated with chronic arthritis in children. N Engl J Med 1985; 313: 1117–1123.

Clarke WL, Shaver KA, Bright GM, et al. Autoimmunity in congenital rubella syndrome. J Pediatr 1984; 104: 370.

Cooper LZ. The history and medical consequences of rubella. Rev Infect Dis 1985; 7(Suppl 1): S1.

Craddock-Watson JE, Ridehalgh MKS, Anderson MJ, et al. Fetal infection resulting from maternal rubella after the first trimester of pregnancy. J Hyg 1980; 85: 381–391.

Cremer NE, Oshiro LS, Weil MK, et al. Isolation of rubella virus from brain in chronic progressive panencephalitis. J Gen Virol 1975; 29: 143–153.

Date M, Gondoh M, Katow S, Fukishima M, Nakamoto N, Kobayashi M, Abe T. A case of rubella encephalitis: rubella virus genome was detected in the cerebrospinal fluid by polymerase chain reaction. No To Hattatsu 1995; 27: 286–290.

Deutman AF, Grizzard WS. Rubella retinopathy and subretinal neovascularization Am J Ophthalmol 1978; 85: 82.

Eckstein M, Vijayalakshmi P, Killedar M, Gilbert C, Foster A. Aetiology of childhood cataract in south India. Br J Ophthalmol 1996; 80(7): 628–632.

Enders G, Nickerl-Pacher U, Miller E, et al. Outcome of confirmed periconeptional maternal rubella. Lancet 1988; I: 1445–1446.

Floret D, Rosenberg D, Hage GN, Monnet P. Hyperthyroidism, diabetes mellitus and the congenital rubella syndrome. Acta Paediatr Scan 1980; 69: 259–261.

Forrest JM, Turnbull FM, Sholler GF, et al. Gregg's congenital rubella patients 60 years later. Med J Aust 2002; 177: 664–667.

Fortuin NJ, Morrow AG, Roberts WC. Late vascular manifestations of the rubella syndrome; a roentgenographic-pathologic study. Am J Med 1971; 51: 134.

Fraser JR, Cunningham AL, Hayes K, Leach R, Lunt R. Rubella arthritis in adults. Isolation of virus, cytology and other aspects of the synovial reaction. Clin Exp Rheumatol 1983; 1(4): 287–293.

Frey TK. Neurological aspects of rubella virus infection. Intervirology 1997; 40: 167–175.

Geiger R, Fink FM, Solder B, Sailer M, Enders G. Persistent rubella infection after erroneous vaccination in an immunocompromised patient with acute lymphoblastic leukemia in remission. J Med Virol 1995; 47(4): 442–444.

Ginsberg-Fellner F, Witt ME, Yahihashi S, et al. Congenital rubella syndrome as a model for type 1 (insulin-dependent) diabetes mellitus: increased prevalence of islet cell surface antibodies. Diabetologia 1984; 27(Suppl): 87–89.

Givens T, Lee DA, Jones T, Ilstrup DM. Congenital rubella syndrome: ophthalmic manifestations and associated systemic disorders. Br J Ophthalmol 1993; 77: 358–363.

Gregg NM. Congenital cataract following German measles in the mother. Trans Ophthalmol Soc Aust 1941; 3: 35–46.

Harcourt GC, Best JM, Banatvala JE, Kennedy LA. HLA antigens and responses to rubella vaccination. J Hyg (Camb) 1979; 83: 412–504.

Hastreiter AR, Joorabchi B, Pujatti G, et al. Cardiovascular lesions associated with congenital rubella. J Pediatr. 1967; 71(1): 59–65.

Heggie AD. Pathogenesis of the rubella exanthem: distribution of rubella virus in the skin during rubella with and without rash. J Infect Dis 1978; 137: 74.

Karounos DG, Wolinsky JS, Thomas JW. Monoclonal antibody to rubella virus capsid protein recognises a B cell antigen. J immunol 1993; 150(7): 3080–3085.

Lawn JE, Ref S, Baffoe-Bonnie B, Adadevoh S, Caul EO, Griffin GE. Unseen blindness, unheard deafness, and unrecorded death and disability: congenital rubella in Kumasi, Ghana. Am J Public Health 2000; 90(10): 1555–1561.

Lim KO, Beal DM, Harvey RL, Myers T, Lane B, Sullivan EV, Faustman WO. Pfefferbaum. A. Brain dysmorphology in adults with congenital rubella plus schizophrenic symptoms. Biol Psychiatry 1995; 37: 764–776.

Lobach DF, Hensley LL, Ho W, Haynes BF. Human cell antigen expression during the early stages of fetal thymic maturation. J Immunol 1985; 135: 1752–1759.

Mauracher CA, Mitchel LA, Tingle AJ. Selective tolerance to the E1 protein of rubella virus in congenital rubella syndrome. J Immunol 1993; 151: 2041–2049.

McIntosh ED, Menser MA. A fifty-year follow-up of congenital rubella. Lancet 1992; 340(8816): 414–415.

Mehl AL, Thomson V. Newborn hearing screening: the great omission. Pediatrics 1998; 101: 1–6.

Menser MA, Forrest JM, Bransbry RD. Rubella infection and diabetes mellitus. Lancet 1978; I: 57–60.

Menser MA, Harley JD, Hertzberg R, et al. Persistence of virus in the lens for 3 years after prenatal rubella. Lancet 1967a; II: 387–388.

Menser MA, Reye RDK. The pathology of congenital rubella; a review written by request. Pathology 1974; 6: 215.

Menser MA, Robertson SEJ, Dorman DC, et al. Renal lesions in congenital rubella. Pediatrics 1967b; 40: 901.

Miller E, Cradock-Watson JE, Pollock TM. Consequences of confirmed maternal rubella at successive stages of pregnancy. Lancet 1982; II: 781–784.

Minor PD, Muir P. Enteroviruses. In: Principles and Practice of Clinical Virology (Zuckerman AJ, Banatvala JE, Pattison JR, Griffiths PD, Schoub B, editors). 5th edition. Chapter 4; 2004; 468–489.

Moriarty BJ. Childhood blindness in Jamaica. Br J Ophthalmol 1988; 72(1): 65–67.

Naeye RL, Blanc W. Pathogenesis of congenital rubella. JAMA 1965; 194: 1277–1283.

Oberfield SE, Cassulo AM, Chiriboga-Klein S, et al. Growth hormone dynamics in congenital rubella syndrome. Brain Dysfunc 1988; 1: 303.

O'Shea S, Best JM, Banatvala JE. A lymphocyte transformation assay for the diagnosis of congenital rubella. J Virol. Methods 1992; 37: 139–148.

Ou D, Jonsen LA, Metzger DL, Tingle AJ. CD4 and CD8 T-cell clones from congenital rubella syndrome patients with IDDM recognise overlapping GAD65 protein epitopes. Hum Immunol 1999; 60: 652–664.

Preece MA, Kearney PJ, Marshall WC. Growth hormone deficiency in congenital rubella. Lancet 1977; 2: 842.

Rand ML, Wright JF. Virus associated idiopathic thrombocytepenic purpura. Transfus Sci 1998; 19(3): 253–259.

Rawls WE, Desmyter J, Melnick JK. Virus carrier cells and virus free cells in fetal rubella. Proc Soc Exp Biol Med 1968; 129: 477–483.

Rawls WE, Melnick JL. Rubella virus carrier cultures derived from congenitally infected infants. J Exp Med 1966; 123: 795–816.

Scasso CA, Bruschini L, Berrettini S, Bruschini P. Progressive sensorineural hearing loss from infectious agents. Acta Ororhinolaryngol Ital 1998; 4(Suppl 59): 51–54.

Sever JL, South MA, Shaver KA. Delayed manifestations of congenital rubella. Rev Infect Dis 1985; 7(Suppl 1): S164.

Smith A. Deafness and Hearing Impairment in Congenital Rubella Syndrome (CRS) Presented at the 5th Meeting of the WHO Steering Committee on Epidemiology and Field Research, Geneva; 4–5 May 1999.

Smith CA, Petty RE, Tingle AJ. Rubella virus and arthritis. Rheum Dis Clin North Am 1987; 13: 265–274.

Steen E, Torp KH. Encephalitis and thrombocytopenic purpura after rubella. Arch Dis Child 1956; 31: 470–473.

Tokugawa K, Ueda K, Fukushige J, Koyanagi T, Hisanaga S. Congenital rubella syndrome and physical growth: a 17-year prospective, longitudinal follow-up in the Ryukyu Islands. Rev Infect Dis 1986; 8(6): 874–883.

Tôndury G, Smith DW. Fetal rubella pathology. J Pediatr 1966; 68: 867–879.

Ueda K, Shingaki Y, Sato T, Tokugawa K, Sasaki H. Hemolytic anemia following post-natally acquired rubella during the 1975–1977 rubella epidemic in Japan. Clin Pediatr 1985; 24: 155–157.

Van Straaten HL, Groote ME, Oudesluys-Murphy AM. Evaluation of an automated auditory brainstem response infant hearing screening method in at risk neonates. Eur J Pediatr 1996; 155(8): 702–705.

Webster WS. Teratogen update: congenital rubella. Teratology 1998; 58(1): 13–23.

Weil MJ, Itabashi J, Creamer NE. Chronic progressive panencephalitis due to rubella virus simulating subacute scherosing panencephalitis. N Engl J Med 1975; 292: 994–998.

Wolinsky JS. Rubella. In: Fields Virology (Fields BN, Poknipe DM, Howley PM, Chanock RM, Melinck J, Monath TP, Roizman B, Straus SE, editors). 3rd edition. Philadelphia: Lippincott-Raven; 1996; pp. 899–929.

Woods WA, Johnson RT, Hostetler DD, Lepow ML, Robbins FC. Immunofluorescent studies on rubella infected tissue cultures and human tissues. J Immunol 1966; 96: 253–260.

Ziring PR, Gallo G, Finegold M, et al. Chronic lymphocytic thyroiditis: identification of rubella virus antigen in the thyroid of a child with congenital rubella. J Pediatr 1977; 90: 419–420.

Rubella Viruses
Jangu Banatvala and Catherine Peckham (Editors)
© 2007 Elsevier B.V. All rights reserved
DOI 10.1016/S0168-7069(06)15003-X

Chapter 3

Laboratory Diagnosis of Rubella and Congenital Rubella

Jennifer M Best[a], Gisela Enders[b]

[a]*King's College London School of Medicine at Guy's, King's College and St Thomas' Hospitals. St Thomas' Hospital, London, SE1 7EH, UK*
[b]*Institute of Virology, Infectiology and Epidemiology e.V and Labor Prof. Dr. med. Gisela Enders and Partner, Rosenbergstraße 85, 70193 Stuttgart, Germany*

Introduction

The accurate laboratory diagnosis of rubella infection in pregnant women presenting with a rubella-like illness, or exposed to possible infection is essential, given that the clinical diagnosis of rubella is unreliable (Chapter 2). Rapid techniques are required to confirm infection in pregnancy so that a woman can be offered a termination of pregnancy (TOP) or prenatal diagnosis (PD) as quickly as possible.

Rubella virus (RV) was first isolated in cell culture in 1962. Although RV grows in many cells, it does not produce a distinct cytopathic effect (CPE) in most cell cultures. Parkman et al. (1962) first isolated RV in primary vervet monkey kidney (VMK) cells, identifying it by interference with an enterovirus (Echovirus 11). In the same year, Weller and Neva (1962) reported isolation of RV in primary human amnion cells in which a CPE was produced. A distinct CPE is also produced in rabbit kidney cells (RK13) and these cells have been the most useful cell line for identification of RV (McCarthy et al., 1963; Best and Banatvala, 1967). Rubella antibodies were first measured by neutralization in VMK cells (Parkman et al., 1962). However, this test was too time consuming and labour intensive for routine use. The development of a haemagglutination inhibition (HAI) test in 1967 was a major advance, since this test was suitable for use in most laboratories and large numbers of sera could be tested within a short time (Halonen et al., 1967; Stewart et al., 1967).

Methods for the detection of rubella IgM were first developed for the rapid diagnosis of rubella in the late 1960s. Initially sera were fractionated on sucrose gradients or by gel filtration and the IgM-containing fractions tested for rubella antibodies by HAI (Vesikari and Vaheri, 1968; Best and Banatvala, 1969). These techniques were somewhat cumbersome and were replaced by radioimmunoassay (Tedder et al., 1982) and later by enzyme immunoassay (EIA) (Hodgson and Morgan-Capner, 1984; Enders and Knotek, 1986; Gerna et al., 1987), which are more rapid and have higher throughput. Detection of rubella-specific IgM remains the method of choice for diagnosis of both postnatally and congenitally acquired rubella, but isolation of virus and detection of viral RNA by nested reverse transcription-polymerase chain reaction (RT-nPCR) are useful for the diagnosis of congenital rubella in the fetus and newborn infant.

Laboratory techniques

Serological methods

There is little antigenic variation among RV strains circulating worldwide, thus, antigen prepared for serological tests with one strain will detect antibodies induced by any RV strain.

Preparation of rubella antigens for use in serological assays

Currently RV antigens used for diagnostic assays are based on whole virus produced in cell cultures. Continuous cell lines, such as BHK-21 and Vero, are best suited for conventional antigen production because of the higher levels of RV produced (Best and O'Shea, 1995). However, even in these cell cultures RV does not grow consistently to high titres, making it difficult to obtain reproducible batches of antigen. This may be overcome by employing recombinant antigens, RV-like particles (RVLPs) and synthetic peptides: definition of optimal antigens has focussed on a region of the immunodominant RV glycoprotein E1, which contains important B-cell epitopes responsible for inducing neutralizing and HAI antibodies (Wolinsky et al., 1993; Starkey et al., 1995). Recombinant antigens produced in *E. coli* (Starkey et al., 1995), via baculovirus–insect cell expression systems (Schmidt et al., 1996; Nedeljkovic et al., 1999), viral-like particles synthesized in transfected cells (Hobman et al., 1994; Grangeot-Keros et al., 1995; Grangeot-Keros and Enders, 1997) and synthetic peptides representing the entire E1 or parts of it (Zrein et al., 1993; Pustowoit and Liebert, 1998; Cordoba et al., 2000; Gießauf et al., 2004) have been assessed for use in EIAs and immunoblotting.

RVLPs expressed in transfected BHK-21 cells carry epitopes for T-cell proliferation and cytotoxicity, as well as B-cell epitopes, as they contain the three RV structural proteins E1, E2 and C. These have been used in a commercial EIA for detection of rubella-specific IgG and IgM. Studies comparing the IgG-EIA with the HAI and two whole virus-EIAs for specific IgG revealed a sensitivity of 94–100% and a specificity of 80–98% (Grangeot-Keros et al., 1995). The sensitivity of the RVLP rubella IgM assay was 100% and the specificity 99.3%, when compared with two other commercial EIAs employing whole virus antigen (Grangeot-Keros and Enders, 1997).

Routine tests

Tests for rubella antibodies and specific IgG

Numerous techniques have been developed for the detection of rubella antibodies (Best and O'Shea, 1995). Their use depends on their practicality and turnaround time. Apart from neutralization, all the tests described below, will provide a result within 24 h, but the appearance of rubella IgG in relation to clinical features will vary according to the test employed (see Fig. 1, p. 20 in Chapter 2).

The haemagglutination inhibition test, which is labour intensive, detects total antibodies and its values are expressed in titres. The lipoproteins in serum are inhibitors of haemagglutination and must be removed from virus preparations used as antigen and from test sera (Halonen et al., 1967; Stewart et al., 1967). Low false-positive results may be obtained if lipoprotein inhibitors are not completely removed from test sera. In the UK and USA, the HAI test is no longer used for rubella antibody screening and diagnosis, although it is still used in Germany and some other European countries.

The enzyme immunoassay is the most common assay used for detection of specific IgG, as it is readily automated and commercial kits are available. The EIA test kit of choice should have a high specificity (> 98%) and a negative predictive value (NPV) of 100% (Pustowoit et al., 1996; Grangeot-Keros and Enders, 1997; L. Hesketh, personal communication, 2003).

The latex agglutination test (LA), which is available commercially, has the advantage of rapidity as it gives qualitative results within a few minutes. Its

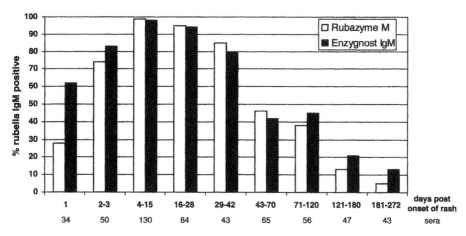

Fig. 1 Detection of rubella IgM following naturally acquired rubella in two commercial tests (552 sera from 238 patients; Enders and Knotek, 1986).

sensitivity compared to the HAI test is 94–98% and it is therefore well suited for screening purposes (Vaananen et al., 1985; Enders, 1987).

Single radial haemolysis (SRH) is less widely used now as it cannot be automated and is subject to observer error. SRH plates are usually prepared in the laboratory using commercially available reagents (Kurtz et al., 1980; Best and O'Shea, 1995). In Germany, the SRH test is commercially available in kit form. Results are available within 24 h. It may be used as a supplementary test for rubella antibody screening to confirm low positive results. In addition, a negative value in the SRH test (diameter < 5 mm) in combination with a low positive HAI result may be indicative of primary acute infection.

Neutralization tests (NT), although labour intensive, are the reference standard for determining "protective" antibodies. Neutralizing antibodies are measured in some reference laboratories to verify immunity, especially when specific IgG levels are low (< 10 IU/ml) following vaccination. The early neutralizing assays were carried out in primary VMK cell cultures employing the interference technique (Parkman et al., 1964; Gitmans et al., 1983). Continuous cell lines such as RK-13 and Vero have also been used, since RV causes subtle CPE and plaque formation in these cells. Tischer and Gerike (2000) showed that the sensitivity of the plaque NT on RK-13 cells was nearly equal to that of EIA because equivocal antibody concentrations in the EIA (4–6 IU/ml) could all be confirmed by plaque neutralization. In these conventional NT, results read by CPE or plaque reduction are available after 9–10 days.

A new NT has been described using Vero cells in microtitre plates. RV-positive cells are stained with monoclonal antibodies (against RV proteins E1, E2, C; Viral Antigens Inc., Tennessee, USA) and horseradish peroxidase-conjugated rabbit anti-mouse antibody, and results are available within 48 h (M. Eggers, G. Enders et al., unpublished data, 2004). Best and O'Shea (1995) have also described a microtitre plate assay using RK13 cells. Most of these NT-assays, which measure the capacity of a given serum to block infection of cells in culture, are technically difficult and time consuming. Other approaches include EIAs based on the synthetic peptides SP15 (Wolinsky et al., 1993; Pustowoit and Liebert, 1998; Cordoba et al., 2000) and BCH-178 (Gießauf et al., 2004). When HAI and SP15-EIA were compared for determining immune status, results demonstrated that SP15-EIA is very specific (100%) and sensitive (98%) for detecting protective antibodies (Cordoba et al., 2000). The BCH-178 EIA detects only a proportion of neutralizing antibodies, as the peptide has insufficient domains for binding neutralizing antibodies (Gießauf et al., 2004).

Detection of rubella IgM antibodies

Detection of rubella IgM is the method of choice for diagnosis of both postnatally acquired and congenitally acquired rubella. Rubella IgM is detected by commercial EIA kits. "Indirect" and "IgM-capture" formats are available. The "indirect" format employs antigen adsorbed to the solid phase and enzyme-labelled anti-human IgM. This type of assay must employ a reagent to remove all rubella-specific IgG, in order to avoid false-positive reactions with sera, which contain rheumatoid factor or other IgM antiglobulins (Meurman et al., 1977; Vejtorp et al., 1979; Enders, 1985). The

"IgM-capture" format employs anti-human IgM attached to the solid phase. Thus, IgG cannot bind and is washed away so that it does not contribute to false-positive reactions. Most IgM-capture assays employ native or recombinant rubella antigen and an enzyme-labelled monoclonal antibody to the antigen (Tedder et al., 1982; Gerna et al., 1987; Grangeot-Keros and Enders, 1997).

Information about development and persistence of specific IgM in the assay employed, together with clinical details, are essential for interpretation of results, including the decision as to whether further sera are required. The sensitivity and specificity of rubella IgM kits must be thoroughly evaluated using a panel of positive and negative sera, sera from patients with other infections and sera-containing autoantibodies (Matter et al., 1994; Hudson and Morgan-Capner, 1996). The performance of seven commercial assays was recently compared using well-defined panels comprising 283 sera from cases of rubella and 417 sera from non-rubella cases (Tipples et al., 2004). Sensitivity of both indirect and IgM capture assays, ranged from 74.1% to 78.9% and specificity from 85.6% to 97.2%. In sera taken < 10 days after onset of rash the sensitivity of the seven assays ranged from 40% to 57.5% (mean 49.8%) and in convalescent phase sera taken ≥ 10 days after onset of rash from 94.3% to 98.8% (mean 97.2%). These results confirmed the experience of earlier studies (Enders et al., 1985; Enders and Knotek, 1986), which include some commercial assays still on the market. They confirmed that all tests perform well at the time of high rubella IgM concentration, but their sensitivity differs when very low rubella IgM concentrations are present, such as shortly after the onset of rash and when concentrations decline several weeks later (Fig. 1).

False-negative IgM results are more likely to occur in indirect assays and may be due to high levels of specific IgG antibodies. The frequency of false-positive results will increase with decreasing positive predictive value when infection becomes rare, e.g. when vaccination campaigns have been effective (Cutts and Brown, 1995). False-positive results may be due to

- cross-reacting IgM antibodies in sera from patients with measles, parvovirus B19, Epstein–Barr virus (EBV) and cytomegalovirus infections;
- polyclonal stimulation through EBV infection;
- rheumatoid factor or other autoantibodies; and
- heat inactivation of sera used in indirect assays.

Supplementary tests

Rubella IgG avidity

Indications for use.   Tests for rubella IgG avidity are useful in the following situations:

- To distinguish primary rubella in early pregnancy from rubella reinfection.

- In cases where rubella IgM has been detected in the absence of a rubella-like illness or contact with a rubella-like illness, to distinguish primary rubella infection from "long-persisting rubella IgM antibodies"(p. 45).

Principles. The avidity or functional affinity of rubella-specific antibodies increases with time after primary exposure to the antigen. Thus, specific IgG is initially of low avidity, but will mature to high avidity within weeks or months after infection depending on the technique used.

Methods to measure IgG avidity are based on EIA. The most common method employs elution of low-avidity antibody from preformed immune complexes of IgG by addition of a denaturating agent to the washing buffer (*elution principle—wash method*). Thomas and Morgan-Capner (1988) have used a different method, adding the denaturating agent to the serum dilution buffer, to prevent the binding of low-avidity antibody to the antigen (*dilution principle*). Several denaturants such as urea (Hedman and Seppälä, 1988; Enders and Knotek, 1989; Hedman and Rousseau, 1989; Eggers et al., 2005) and diethylamine (DEA) (Thomas and Morgan-Capner, 1988, 1991; Thomas et al., 1992a,b; Böttiger and Jensen, 1997; Enders, 2005) have been used for this purpose. Results of avidity tests are usually expressed as percent ratio (Enders and Knotek, 1989; Böttiger and Jensen, 1997; Eggers et al., 2005) or as ratio of EIA absorbance values of a single serum dilution with and without denaturation (= *index calculation method*) (Pustowoit and Liebert, 1998). When using the DEA *shift value method* (DSV) (Thomas and Morgan-Capner, 1988, 1991) results are expressed as distance between two curves drawn from absorbances of serial sera dilutions with and without the protein denaturant.

Urea wash method. The urea wash method (elution principle) has been used to measure avidity after primary infection (Hedman and Seppälä, 1988; Enders and Knotek, 1989; Hedman and Rousseau, 1989), rubella vaccination (Enders and Knotek, 1989) and rubella reinfection (Enders and Knotek, 1989; Hedman and Rousseau, 1989). An enzyme-linked fluorescent assay (ELFA) using 6 M urea will soon be commercially available (VIDAS Rubella Avidity). A recent evaluation has shown that an avidity index greater than 40% is indicative of high avidity and allows the exclusion of recent natural infection or vaccination < 3–5 months before serum collection (Eggers et al., 2005).

DEA shift value method. This method (dilution principle) has been used extensively by Thomas and Morgan-Capner (Thomas and Morgan-Capner, 1988, 1991; Thomas et al., 1992a,b), who have also compared available methods to define optimal avidity testing conditions for differentiation between primary rubella, early postvaccination, natural or vaccine infection in the remote past, reinfection and "persistent rubella-specific IgM reactivity" (Thomas et al., 1992a,b).

DEA wash method. The DEA wash method (elution principle) of Böttiger and Jensen (1997) employs 35 mM DEA and can be used with the Enzygnost anti-rubella-IgG-EIA (Dade Behring). It is important that patients' sera are prediluted to give a rubella IgG concentration of 70 IU/ml, because the rise of the avidity index may be detected earlier and at incorrectly high values if there is a high specific IgG concentration (G. Enders, unpublished results).

The kinetics of avidity maturation following primary infection and vaccination of seronegative women are shown in Fig. 2. In patients with *primary rubella* there was a rise from low (<30%) to moderate (30–50%) avidity 6–12 weeks after onset of rash and progression to high avidity (>50%) after about 4–5 months. Between 6 and 12 months after infection 90–100% of patients had high-avidity antibody, which has been shown to persist for many years. Avidity rises more slowly after *rubella vaccination with* high-avidity antibody detected in <10% of females 5 months after vaccination, in 20–40% at 5–9 months and in 50% at 10–12 months. In approximately 30% of vaccinees avidity will remain at moderate levels for many years.

When differentiating *reinfection* from *primary rubella* in pregnancy with the DEA wash method, maturation of avidity is faster following reinfection (2–4 weeks after rubella contact) than following a primary infection (about 4–5 months following primary rubella).

The urea wash method has been extensively used, but comparative studies have shown the DSV and DEA wash methods will detect low-avidity antibody for longer after primary infection than the urea wash method (Table 1), thus prolonging the time available for diagnosis of naturally acquired rubella. As the DEA wash method is easier to perform than the DSV method, it is now the method of choice.

IgG avidity and long-persisting rubella-specific IgM. Rubella IgM antibodies may persist for as long as 6 years and may be difficult to distinguish from IgM produced following primary rubella and rubella reinfection (Enders, 1991; Thomas et al., 1992b; Böttiger and Jensen, 1997; Enders, 1997, 2005). If sera are obtained within the first 12–16 weeks of pregnancy, then avidity tests are useful to distinguish IgM antibodies due to *primary rubella* from *long-persisting IgM antibodies*. In those with long-persisting IgM, the avidity indices will be moderate (mainly in vaccinees) or high (mainly post natural infection) irrespective of HAI and rubella IgG antibody levels. This contrasts with primary infections where the rise from moderate-to-high avidity occurs 4–5 months after onset of infection. However, confirmation of long-persisting IgM comes from testing consecutive serum samples for 8 months or more after their initial detection.

By testing more than 448 mother–infant pairs, it was shown that correctly defined long-persisting rubella IgM antibodies in the mother were not a risk for congenital infection (Enders, 2005). Rubella IgM antibodies were found at the same concentrations in the mothers at birth and during repeated testing in pregnancy, while rubella IgM was not detected in newborn sera, thereby excluding congenital infection.

Non-reducing immunoblot (IB)

Non-reducing IB assays were developed for detection of rubella-specific immunoglobulins directed to topographical epitopes on the RV structural proteins (Zhang

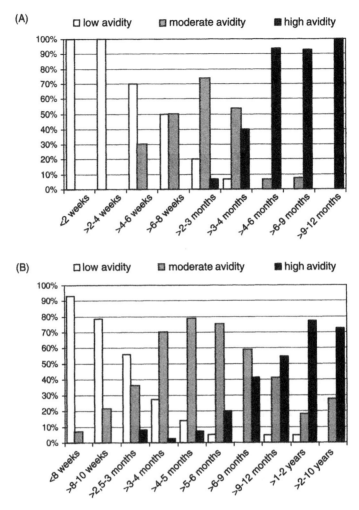

Fig. 2 Distribution of rubella-specific IgG avidity at various time points in (A) patients with primary rubella infection (203 sera from 105 patients) and in (B) previously seronegative rubella vaccinees (278 sera from 159 vaccinees, most of whom were vaccinated with RA27/3) (modified DEA wash method; G. Enders, unpublished results, 2004).

et al., 1992). With a commercial non-reducing IgG IB (recomBlot Rubella IgG, MIKROGEN GmbH, Martinsried, Germany) based on recombinant rubella antigens (Hobman et al., 1994), it is possible to visualize IgG antibody reaction to the RV polypeptides C, E1, E2 and the dimer E1E2 by specific bands (Fig. 3). During the first few days after onset of rash, nearly all sera develop IgG antibodies to C and E1, but the E2 response is delayed (Meitsch et al., 1997). E2 antibodies appear approximately 3–4 months after primary infection and approximately 5 months after vaccination. In contrast to recent primary infection where antibodies to the E2

Table 1

Percentage of sera with low-avidity rubella IgG_1; comparison of three methods using 206 sera from 171 patients with primary rubella (data from Fig. 1 of Thomas et al., 1992a)

Method	Time after onset of acute rubella infection							
	≤ 14 days	15–28 days	≥ 1–2 months	≥ 2–3 months	≥ 3–4 months	≥ 4–5 months	≥ 5–7 months	> 7 months
DSV (%)	100	97	94	78	91	71	21	0
DEA wash (%)	~98	~95	~75	~44	~38	0		
Urea wash (%)	~80	~62	~24	~10	0			

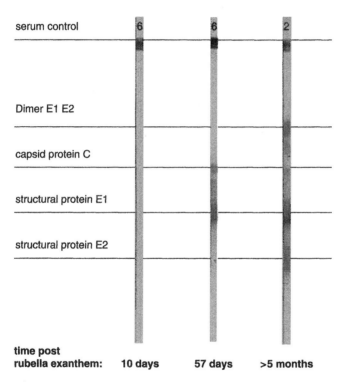

Fig. 3 Kinetics of the IgG response to the rubella polypeptides C, E1 and E2 in consecutive sera from a patient with rubella infection, visualized by specific bands in a commercial non-reducing IgG immunoblot (recomBlot Rubella, MIKROGEN).

band develop in 90% (5–6 months after infection), <60% of vaccinees develop antibodies to E2 even 2–10 years after vaccination (Fig. 4).

The development of IgG antibodies against the E2 protein has proved a useful additional assay for differentiating reinfection from primary infection, since in

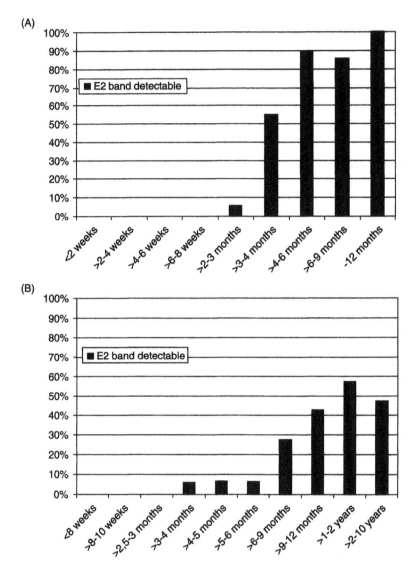

Fig. 4 Immunoblot analysis of the IgG response to recombinant E2 protein in (A) patients with primary acute rubella infection (255 sera of 129 patients) and in (B) previously seronegative rubella vaccinees (269 sera of 168 vaccinees, of whom most were vaccinated with RA27/3) (non-reducing IgG immunoblot: recomBlot Rubella, MIKROGEN; G. Enders, unpublished results, 2004).

reinfection the E2 band is present within about 4 weeks after contact, in contrast to 3 months after primary infection. Non-specific binding may occur in about 8% of sera (G. Enders, unpublished results). In patients with cow's milk allergy, anti-bodies to bovine proteins may react with milk protein used in the IB-buffers.

Alternative specimens for diagnosis and surveillance

Oral fluid.   Oral fluid consists of secretions from the salivary glands and crevicular fluid. It can be used for detection of virus-specific antibodies, since IgG and IgM antibodies reflect those in the serum (Mortimer and Parry, 1991; McKie et al., 2002) as well as for detection of virus. Oral fluid can be easily collected from the mouth using an absorbent device (Fig. 5); several such devices are available commercially (Vyse et al., 2001).

Since the collection of oral fluid is a non-invasive technique, it is particularly useful for children. In the UK, oral fluid tests are used to investigate rash illness in populations where measles and rubella are rare, due to MMR vaccination programmes. In a recent UK study, none of the 195 children (aged under 16 years) with a rash illness had measles or rubella IgM, and other infections, including parvovirus B19, group A streptococcus and human herpesvirus-6 were confirmed in 48% of these children (Ramsay et al., 2002).

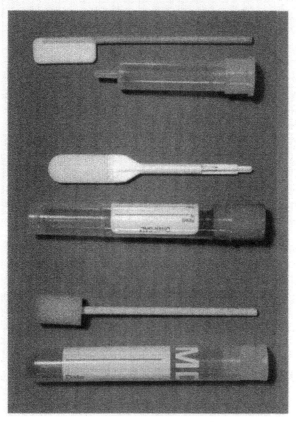

Fig. 5 Oral fluid collection devices. From top to bottom: Orasure, Omni-SAL, Oracol (Reprinted from Vyse et al. 2001 with permission from the author and the Royal Institute of Public Health).

Oral fluid is particularly useful for seroepidemiological studies, as it is easier to collect and a phlebotomist and sterile equipment including needles are not required (Mortimer and Parry, 1991; Nokes et al., 1998). In developing countries where needles and syringes are often reused, this reduces the risk of transmitting blood-borne infections. Methods for the extraction, transportation and preservation of samples have been established (Mortimer and Parry, 1991; Nokes et al., 1998; Morris et al., 2002).

Sensitive assays are required to detect specific antibodies in oral fluid, as antibody concentrations are much lower than in serum. Consequently antibody capture assays are preferred for detection of specific IgG and IgM. Rubella-specific IgG or IgM antibody capture radioimmunoassays (GACRIA or MACRIA) were used initially, but the IgG assay has since been adapted to amplification-based EIA (Vyse et al., 1999). It is possible to use a modified commercial assay with antigen on the solid phase to measure rubella-specific IgG in oral fluid (Ben Saleh et al., 2003). Detection of rubella-specific IgG and IgM antibodies in oral fluid compare well with serum (Perry et al., 1993; Ramsay et al., 1998; de Oliveira et al., 2000). Oral fluid tests are more efficient at detecting antibodies in children than in adults (Nokes et al., 1998); they are not sufficiently sensitive and specific for rubella susceptibility screening of adults. In order to detect rubella IgM, oral fluid should be collected between 1 and 5 weeks after onset of rash. Oral fluid can also be used for the diagnosis of congenital rubella (Eckstein et al., 1996), and the measurement of IgG avidity (Akingbade et al., 2003). The widespread use of oral fluid tests for rubella awaits the development of suitable commercial assays, such as those recently developed for measles and mumps.

Using RT-nested PCR (RT-nPCR; p. 52) it is possible to detect RV RNA in oral fluid collected with the Oracol device (Fig. 5; Malvern Medical Developments, Worcester, UK), for which oral fluid should be collected within 7 days after onset of rash (Jin et al., 2002). This method is also useful for the diagnosis of congenital rubella and can be used to obtain RV RNA for sequencing for molecular epidemiological purposes (Jin et al., 2002; Vyse and Jin, 2002; Cooray et al., 2006).

Dried blood spots (DBS).   DBS have also been used successfully for detection of rubella IgM and IgG antibodies (Helfand et al., 2001; Karapanagiotidis et al., 2005) and viral RNA (Pitcovski et al., 1999; De Swart et al., 2001). DBS are collected by finger prick onto high-quality filter paper (e.g. Schleicher and Schuell, 903). Guthrie cards can also be used. They can be stored at room temperature (Parker and Cubitt, 1999). Packing and transportation of DBS is easier than sera, as they are not considered as "Dangerous Goods".

Virological methods

Detection of RV is seldom used for diagnosing postnatal rubella, since serological methods are quick and reliable. RT-nPCR is preferred for PD and detection of RV in congenital rubella cases due to its greater sensitivity and rapidity (see below).

However, virus isolation from clinical specimens is helpful to obtain sufficient RV for sequencing, e.g. in order to carry out phylogenetic analysis, and to differentiate wild-type and vaccine virus (Bosma et al., 1996; Frey et al., 1998; Zheng et al., 2003; World Health Organisation, 2005; Cooray et al., 2006).

Virus isolation

This procedure is time-consuming and labour-intensive and only carried out in specialized laboratories. Results may not be available for up to 4 weeks. Specimens should have two passages in Vero cells, with subsequent identification of RV in RK13 cells, in which a distinct CPE is usually produced (Fig. 6).

Alternatively RV may be identified in the cell cultures by indirect immunofluorescence using hyperimmune or monoclonal antibodies (Best and O'Shea, 1995). Punctate staining throughout the cytoplasm occurs with antibodies to the C protein and intense staining of the Golgi apparatus with antibodies to the E2 and E1 proteins. Alternatively RV-positive cells can be identified by RT-nPCR (Revello et al., 1997). Vero/SLAM cells can be used for the isolation of both measles virus and RV. Thus, it has been suggested that measles virus-negative cell cultures, could be used for the detection of RV by indirect immunofluorescence (Best et al., 2005).

Recently a novel cell-culture-based assay for RV (genotypes I and II) detection in clinical specimens was described, which combines cell culture techniques, the use of genetically modified RV genes and reporter gene expression (Tzeng et al., 2005). Preliminary evaluation demonstrated good sensitivity and specificity, but this method may not be suitable for routine use.

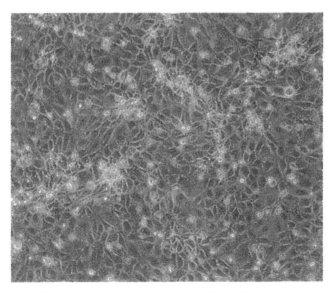

Fig. 6 Cytopathic effect caused by rubella virus in RK13 cells 3 days after inoculation.

Reverse transcription-polymerase chain reaction

The high GC content of the RV genome has made the development of a reliable RT-PCR difficult and a variety of protocols were validated to find the most suitable for different purposes. Diagnostic applications include invasive PD and diagnosis of congenital rubella (pp. 58, 66). Highly sensitive PCR assays, detecting 3–10 copies of RV RNA, are required for PD as fetal specimens generally contain low RV copy numbers. The RT-nPCR protocol described by Bosma et al. (1995a,b) has been used fairly widely. This RT-nPCR amplifies a 143 bp sequence from a well-conserved region of the E1 gene and has a sensitivity of ≤ 2 RNA copies. Other RT-nPCR protocols targeting the E1 gene have been used successfully for PD (Revello et al., 1997; Macé et al., 2004). Revello et al. (1997) compared direct RT-nPCR, culture-RT-nPCR and virus isolation and found sensitivities of 100%, 75% and 75%, respectively. Studies have also shown that RT-nPCR is about 20% more sensitive than virus isolation for PD. A high reproducibility of RT-nPCR results with fetal specimens stored for 1–4 years at $-70°C$ has been observed using the RT-nPCR of Bosma et al. (1995a,b).

Specimens should be transported at 4°C or on dry ice or liquid nitrogen if in transit for > 12 h, as viral RNA is easily destroyed at ambient temperatures. To avoid false-negative results, several strategies, including spiking and non-competitive internal controls, have been established to monitor PCR inhibitors. For PD it is especially important to check the quality of nucleic acid extraction, because the PCR efficiency can be dramatically reduced by the presence of PCR-inhibitory substances (e.g. haem or added anticoagulants such as heparin). Non-competitive internal controls such as keratin mRNA (Bosma et al., 1995a) are widely used, however, these endogenous sequences may vary from one sample to another and underestimate the amount of PCR inhibitor in the sample. For this reason a competitive PCR, in which a synthetic construct is used, is generally favoured. This is amplified by the same primers, but detected by a different molecular probe and is able to monitor both the quality of nucleic acid extraction and PCR (Revello et al., 1997).

A longer amplicon is required for genotyping of RV isolates for molecular epidemiological studies. A RT-nPCR, which amplifies a 592-bp sequence from the E1 gene has been described for this purpose (Cooray et al., 2006). A multiplex nPCR to detect measles virus, RV and parvovirus B19 with the detection limit of 10 genome equivalents may be useful in countries where these viruses are circulating (del Mar Mosquera et al., 2002). Viral RNA is stable on certain filter papers for later investigation by RT-PCR, which is useful for specimens sent to distant laboratories.

Facilities for rubella testing and their distribution

In countries, such as UK, USA, Finland and Sweden, where there is currently little or no rubella circulation and limited demand for diagnosis, most hospital/local laboratories provide tests for rubella antibody screening (including confirmatory assays) and rubella IgM. Alternative techniques for rubella IgM, IgG and IgG avidity assays, IB, virus isolation and tests for PD should be available in reference laboratories, which should also be responsible for evaluating available assays. In Germany, France and North Italy such additional tests are usually performed by national reference and university laboratories as well as by private expert laboratories. Specialized tests are required for use with oral fluid.

A global measles–rubella laboratory network has been established by WHO, which comprises global specialized laboratories, regional reference laboratories, national and subnational laboratories (Robertson et al., 2003). Laboratory support, training and quality control are available for this network. In addition to conducting serological assays, these laboratories are encouraged to obtain RV isolates for molecular epidemiological studies (Best et al., 2005).

Rubella antibody screening

In the UK, rubella antibody screening is offered to all pregnant women and to women with no history of rubella or MMR immunization or a previous positive rubella antibody test. In Germany, rubella antenatal screening is obligatory in early pregnancy since 1972. The aim of the antenatal screening programmes is to identify women who are susceptible to rubella, for whom vaccine is advised postdelivery and not to identify or exclude rubella infection in the current pregnancy. It is particularly important that screening is offered to women who come from countries where they may not have been offered rubella vaccination (Tookey et al., 2002; Department of Health, 2003; Robert-Koch-Institut, 2004). In the UK, testing is considered unnecessary if there is documented previous evidence of two reliable tests on different serum samples both confirming the presence of rubella-specific IgG (Department of Health, 2003). However, most UK clinics may find it more convenient to test all antenatal patients regardless of past history of testing or vaccination. In Germany, antenatal antibody testing is considered unnecessary if there is documented HAI titre of $\geq 1:32$ or an IgG antibody level $\geq 15\,\text{IU/ml}$ before the current pregnancy. Apart from that, all women should be tested prior to *in vitro* fertilization, irrespective of previous test results or vaccination history.

In the UK, many other European countries and the USA, the main tests used for antibody screening are commercially available EIAs and LA and SRH may be used as second-line/confirmatory assays. In contrast, in Germany the HAI test assisted by IgG EIAs is for the time being the main screening test. The cut-off for rubella antibody detection is currently $10\,\text{IU/ml}$ (Skendzel, 1996; Department of Health, 2003), or an HAI titre of $1:32$ in Germany. Sera giving low levels ($< 10\,\text{IU/ml}$, or HAI titre $1:8/1:16$) or equivocal results on initial testing should be re-tested using an alternative assay which may also help to monitor the reliability of the first assay (O'Shea et al., 1999). In the UK, screening results are reported as "rubella antibody detected/not detected" rather than "immune/susceptible" (Department of Health, 2003). The presence of rubella-specific IgG does not exclude recent infection and if a woman has a history of rash or contact with a rash further tests are required.

If a low level ($<10\,IU/ml$, or HAI titre $1:8/1:16$) of rubella antibody is detected by two reliable assays in a woman with a documented history of two or more rubella vaccinations, further doses of vaccine are unlikely to be of value as persistent rises in antibody concentration are seldom achieved (Matter et al., 1997; Tischer and Gerike, 2000), and protection against rubella and congenital rubella is assumed. Nevertheless such women are advised to report any rash illness or contact with a rubella-like rash, so that further investigations can be carried out (Morgan-Capner and Crowcroft, 2002).

Diagnosis of rubella

Postnatally acquired rubella

Serological investigation for postnatally acquired rubella, especially in pregnancy, is essential to confirm or refute rubella infection. Measles, enteroviruses, HHV-6 and HHV-7, group A streptococcus and dengue virus may present with a similar rash, and parvovirus B19, chikungunya-, Ross River-, West Nile- and Sindbis viruses may present with both rash and joint symptoms (Banatvala and Brown, 2004).

Rubella is usually diagnosed serologically. Detection of rubella IgM in a serum taken 3–6 days after onset of rash is the method of choice for diagnosis of acute rubella. Although specific IgM may be detected earlier in some patients, this will depend on the assay used (p. 43). If no specific IgM is detected in a serum taken <6 days after onset, it is advised to collect a second serum a few days later. This should be tested in parallel with the first serum for specific IgM and IgG, as a significant rise in rubella IgG concentration may be detected in addition to rubella IgM. Rubella IgG antibodies are usually detectable by well-controlled commercial EIAs within 7 days after onset of rash while HAI antibodies may be detected within 2 days (see Chapter 2, Fig. 1, p. 20).

If rubella-specific IgM is detected in a patient with rash, this is likely to indicate recent primary rubella, but care is needed in interpretation (pp. 42–43). The patient's history and results of any previous tests should be considered and supplementary tests, previously described, may be required to confirm the diagnosis (Best et al., 2002).

Diagnosis of women with rubella-like illness in pregnancy

The risk of congenital rubella syndrome (CRS) is greatest in the first 12 weeks of pregnancy (Chapter 2). Therefore, it is particularly important to test women who are in the first trimester, rapidly, so that if they wish to consider a pregnancy termination following a diagnosis of rubella, this can be done as early as possible. Close collaboration between clinical and laboratory staff is essential if serological tests are to be interpreted accurately. In order to interpret laboratory results accurately, the following information is required:

* date of last menstrual period (LMP);
* previous documented history of rubella antibody screening and rubella/MMR vaccination;
* date and duration of contact;
* type of contact (e.g. significant: household or at work);
* age of index case; and
* symptoms or rash and rubella/MMR vaccination history in the index case.

Serum should be collected as soon as possible after onset of symptoms and tested for rubella-specific IgG and IgM antibodies. A second serum is usually required 5–10 days later to demonstrate seroconversion and to confirm the presence of IgM, if this was detected in the first sample. Accurate information on the date of onset of illness and the distribution of rash, lymphadenopathy and arthropathy (Chapter 2), should be obtained in order to interpret serological results (see Fig. 1, p. 20).

The results on sera from women who present more than 4–6 weeks after onset of symptoms may be difficult to interpret, as specific IgM may have become undetectable in about 20% of cases depending in part on the sensitivity of the IgM assay used. The diagnosis in such cases may be clarified by applying the IgG avidity and IB tests to sera obtained in the first 12–16 weeks of gestation. In sera taken in this time period, moderate-to-high IgG avidity together with the E2 band on IB confirms that IgG and IgM antibodies have not been acquired by recent primary infection in the current pregnancy (Enders, 2005).

Assessment of women exposed to rubella-like illnesses in pregnancy

UK Public Health Laboratory Service joint working party of the advisory committees of virology and vaccines and immunization has produced guidelines on the management of rash illness in pregnancy (Morgan-Capner and Crowcroft, 2002). Women who have been exposed to a rubella-like illness are more likely to acquire infection if exposure is close and prolonged, such as within their own household or at work. Thus, in populations without successful childhood vaccination programmes, children may acquire infection at school and present a risk to a seronegative mother. Women who have had only a brief exposure may usually be reassured initially, although testing is still recommended.

A documented history of two previous positive IgG antibody tests or two documented doses of rubella vaccine or one documented dose of vaccine followed by a positive IgG antibody result, indicates past rubella infection or vaccination (Best et al., 1989). Although further rubella testing seems unnecessary, in practice a woman with significant exposure would be tested. Seronegative women and/or those with low levels of antibody (< 10 IU/ml) should be retested at 7–10 day intervals for up to 4 weeks after contact. When the pregnant woman is seronegative, it may be desirable

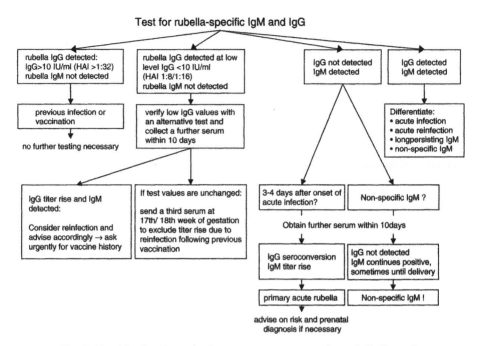

Fig. 7 Algorithm for diagnosis of pregnant women exposed to rubella-like rash.

to collect serum or oral fluid from the index case, in order to confirm or refute the diagnosis of rubella. If the index case is found to be seronegative further follow-up of the pregnant woman is unnecessary, but she should be vaccinated postpartum.

It is always useful to test stored serum samples from the current or a previous pregnancy, as such a serum can be tested in parallel with the later serum sample (Morgan-Capner and Crowcroft, 2002). This is possible in the UK and some other countries where it is recommended to store the antenatal screening sera at −20°C for at least 1 year (Department of Health, 2003). If there is no documented information and also no stored serum sample available, the advice given in the algorithm (Fig. 7) should be followed. However, it may be easier and quicker to take advantage of the early use of supplementary tests.

Diagnosis of rubella reinfection in pregnancy

Rubella reinfection is defined as a rubella-specific immune response in someone with previous naturally acquired or vaccine-induced rubella. Reinfection is usually subclinical and occurs mainly in persons with vaccine-induced immunity who are closely exposed to rubella. Maternal rubella reinfection and its possible risk of fetal infection and damage was an important diagnostic issue in such countries as the UK, USA, Germany and Israel in the late 1980s and 1990s when rubella continued to circulate among children (Bullens et al., 2000). Since then the risk of exposure of pregnant women to circulating RV has decreased due to increased vaccination coverage in these countries. The risk of congenital rubella defects after subclinical maternal reinfection in the first 12 weeks of pregnancy has not been precisely determined. It has been estimated to be <10% and probably <5% (Morgan-Capner et al., 1991; Morgan-Capner and Crowcroft, 2002). Congenital rubella defects have not been reported after reinfection occurring beyond 12 weeks' gestation. Thus, it is important to distinguish rubella reinfection from primary infection in the first 12 weeks of pregnancy, as the risk of fetal damage after reinfection is considerably lower than after primary infection (>80%). Invasive PD is useful to assess whether fetal infection has occurred. Although symptomatic maternal reinfection is very rare, it is considered to pose a higher risk to the fetus compared to asymptomatic maternal reinfection (Morgan-Capner and Crowcroft, 2002), as observed in symptomatic and asymptomatic primary rubella infection (Miller et al., 1982).

Maternal reinfection is diagnosed by a significant rise in rubella IgG and/or HAI concentration, sometimes to very high levels, in a woman with pre-existing antibodies (Best et al., 1989; Enders and Knotek, 1989; Morgan-Capner et al., 1991). Rubella IgM is usually detected if sera are collected within 4–6 weeks of contact. Ideally, pre-existing antibodies should be confirmed in a stored serum sample. However, if such a serum is not available, evidence of pre-existing antibody may be accepted if there are at least two previous laboratory reports of antibodies ≥10 IU/ml obtained by reliable techniques, or a single serum with rubella antibodies ≥10 IU/ml obtained after documented rubella vaccination (Best et al., 1989). Supplementary tests like IgG avidity and immunoblotting may be used to distinguish reinfection from primary infection (pp. 44–47). It is less safe to exclude primary rubella in early pregnancy if serum is obtained beyond the 16th week of gestation. Characteristic antibody responses seen in reinfection are shown in Fig. 8.

Diagnosis of congenitally acquired rubella

Prenatal diagnosis

The value of invasive PD is limited by the fact that virological and serological investigations have to be delayed until relatively late in gestation (21–24 weeks), and for 7–8 weeks after the onset of the rubella-like illness. Furthermore, the number of studies in which sensitive techniques have been employed to detect congenital rubella infection (Table 2) are relatively small and few have reported the outcome of pregnancy, which is necessary for calculating sensitivity, specificity, positive and negative predictive values of available techniques. Thus, it is difficult to compare the reliability of investigations employing different specimens from the conceptus. Additionally, the duration of pregnancy at which TOP may be carried out varies from country to country.

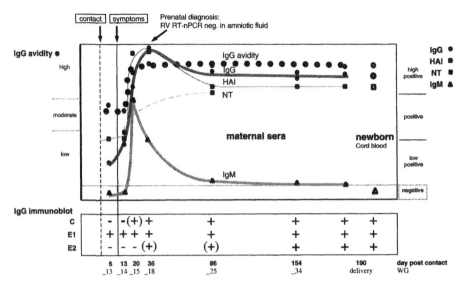

Fig. 8 Schematic antibody pattern for routine and supplementary serologic assays in confirmed rubella reinfection. Results from a case of confirmed symptomatic rubella reinfection in the 13th week of pregnancy following previous vaccination at age 19 and exposure to own child with clinical and serologically confirmed rubella. The baby had no evidence of congenital infection and was healthy at birth and at age 5. IgG immunoblot: +, -antibody is detectable; (+) possibly detectable; –, not detectable; WG, week of gestation.

No official guidelines exist for applying invasive PD to the diagnosis of congenital rubella infection. However, these techniques have been in use for 15 years and are of value in the following cases:

- Laboratory confirmed primary rubella in the early second trimester (12–18 weeks gestation).
- When primary rubella in the first trimester cannot be excluded due to equivocal rubella IgM results, despite the use of supplementary tests (pp. 44–47).
- Laboratory confirmed rubella reinfection before 12 weeks' gestation.

Invasive PD is NO longer required in the following situations:

- Laboratory confirmed primary rubella before the 12th week of gestation, since the high risk of fetal infection and damage is well established (Chapter 2) and the majority of parents are disinclined to continue with the pregnancy.
- When a seronegative pregnant woman has been inadvertently vaccinated with a rubella-containing vaccine, as this has not been associated with symptomatic CRS (Chapter 4).
- Long-persisting IgM antibodies are detected in pregnant women, since they are not associated with congenital infection (pp. 45–46).

Table 2

Results of prenatal diagnosis after acute rubella infection in early pregnancy by RT-nPCR: confirmed/not confirmed by outcome of pregnancy

Study	Pregnant women		RV RNA detected				RV RNA not detected			
	n^a	With symptom	n^a	Confirmed[b]	Not confirmed[b]	Lost to follow-up	n^a	Confirmed[b]	Not confirmed[b]	Lost to follow-up
Bosma et al. (1995b), UK	7		4	1	1	2	3	—	1	2
Tanemura et al. (1996), Japan	34	21	8	5	—	3	26	20	4	2
Revello et al. (1997), Italy (retrospective)	8	8	8	6	—	2	0	—	—	—
Katow (1998), Japan	253	112	48	22	4	22	205	130	13	62
Henquell et al. (1999), France	6	6	1	1	—	—	5	2	—	3
Macé et al. (2004), France	45	20	20	11	—	9	25	6	1	18
Enders et al. (1994–2004), Germany	77	77	58	33	1	24	19	8	1	10

[a]Number of cases tested.
[b]*Criteria of congenital infection:* termination of pregnancy: RV RNA and/or RV detection in fetal tissue; live-born infant: CRS and specific IgM antibody detection at birth; and persistence of specific IgG antibodies until 10 months of age.

Invasive PD is usually accompanied by ultrasound evaluation. However, in the case of maternal rubella, diagnosis of fetal infection on the basis of ultrasound findings has not been reported, even though certain fetal abnormalities (e.g. cerebral ventriculomegaly with intracranial calcifications, cardiac malformations, microcephaly) could potentially be detected by ultrasound (Crino, 1999; Enders et al., unpublished data, 2006). The failure to detect abnormalities by ultrasound is mainly due to the fact that pregnancies with acute rubella up to the 12th week of gestation are mainly terminated before the above-mentioned ultrasound abnormalities become visible beyond the 20th week of gestation.

There seems to be little point in conducting invasive PD if TOP is unacceptable to the parents. However, in such cases, a negative PD result will provide some reassurance to the mother during the inevitably anxious time. Before performing PD it is essential that the risks of the sampling techniques as well as the limited value of positive PD results for prognosis of damage in the newborn are explained to the mother by those responsible for her care. It should be clarified that in case of positive PD results, the risk of fetal damage depends mainly on the gestational week of maternal infection.

Fetal samples used for prenatal diagnosis

Chorionic villus biopsies, amniotic fluid and fetal blood (Table 3) have been used for detection of RV RNA by RT-nPCR, which can provide a result within 48 hours (p. 52). Fetal blood can also be used for the detection of rubella IgM (p. 42).

- *Amniotic fluid* is considered the optimal sample for PD as amniocentesis is less invasive than fetal blood sampling and amniotic fluid can be obtained in larger amounts than fetal blood. Ideally, amniotic fluid samples should be taken 7–8 weeks after rubella infection and after the 21st week of gestation (Enders et al., 2001; Macé et al., 2004).
- *Chorionic villus biopsies* have not been used as extensively as amniotic fluid as detection of RV may not reflect fetal infection as infection of the villi is likely to occur before infection of fetal tissues. A positive result may also be due to RV in contaminating maternal blood or decidual tissue. However, this can be clarified by DNA typing with short tandem repeats for differentiation of maternal and fetal tissue or blood. The optimal time for collecting chorionic villus biopsies is not known, but collection should not be performed before 11th week of gestation on account of the risk of limb-reduction-defects (Jauniaux and Rodeck, 1995). RV RNA positive results have been obtained 2–10 weeks after onset of maternal rash. When a negative result is obtained from chorionic villi, amniotic fluid or fetal blood should be obtained later in pregnancy to confirm this result.
- *Fetal blood*, taken by cordocentesis, can be used for detection of RV by RT-nPCR as well as for detection of rubella IgM. It should be collected 7–8 weeks after onset of maternal rubella and not before the 21st week of gestation.

Table 3

Fetal samples for the prenatal diagnosis of congenital rubella infection

Sample	Optimal time for sampling (weeks of gestation)	Detection of RV RNA by RT-nPCR[a]		Technical risk of fetal loss	Comments
		Sensitivity	Specificity		
Amniotic fluid	≥ 21	86% (31/36)[b]	91% (10/11)[c]	0.2–1%	Sampling possible after 15 weeks, but virus more likely to be detected at 21 weeks or more
Chorionic villus biopsy	11–16	83% (5/6)[b]	Insufficient data	0.2–3%	Positive result may be due to contamination with maternal blood or decidual tissue
Fetal blood	≥ 21	73% (16/22)[b]	100% (7/7)[c]	0.5–2%	Collection is technically more demanding. Can also be used for detection of rubella IgM (see Table 4)

[a]Unpublished results from G. Enders and colleagues.
[b]Number of true positives/total number of fetuses or newborns congenital infected.
[c]Number of true negatives/total number of fetuses or newborns congenital uninfected.

Although IgM antibodies may be produced by the fetus as early as the 12th week of gestation, rubella IgM levels will be too low to be detected before 21 weeks (Enders and Jonatha, 1987; Jacquemard et al., 1995; Enders, 2005). Heparin should not be used when collecting fetal blood samples, as it may inhibit the PCR reaction.

All specimens should be tested for RV RNA by RT-nPCR in triplicate at least, and two or three assays should ideally be used for IgM testing. Observing these recommendations will reduce false-negative RT-nPCR and rubella IgM results. False-negative RT-nPCR results may be due to the fact that virus may not be present in fetal samples at detectable concentrations at the time of sampling.

Practical considerations

When results are available, the virologist should discuss them with the obstetrician and parents, in order to help the mother decide whether to continue with the pregnancy. From the few published studies of PD, there is insufficient information available on the reliability of RT-nPCR and rubella IgM detection in fetal samples. Results should therefore be interpreted with caution. Ultrasound beyond the 22nd week of gestation may be helpful in excluding major defects, but it cannot be used for affirming fetal infection.

Seven studies of invasive PD of congenital rubella are listed in Table 2. Henquell et al. (1999) and Macé et al. (2004) tested only amniotic fluid and fetal blood, but in the other five studies amniotic fluid, chorionic villus biopsies and fetal blood samples were included. In the study by Enders et al. results were also assessed by pregnancy outcome (Table 4). In this study, a case was designated as congenital rubella infection if one fetal sample (chorionic villus biopsy, amniotic fluid or fetal blood) was RV RNA positive by RT-nPCR, or rubella IgM was detected in fetal blood. Ninety-seven

Table 4

Prenatal diagnosis in cases with confirmed primary rubella infection: Detection rate for RV RNA using RT-nPCR (E1 primers) in various fetal samples and/or for rubella-specific IgM in fetal blood and calculation of sensitivity, specificity and predictive value based on the outcome of pregnancy (Enders et al., unpublished results)

No. positive/total no. tested (%)	Cases	
	RT-nPCR[a]	Rubella IgM[a]
	58–77 (75)	32–46 (70)
With known outcome (*n*)	43	29
Sensitivity[b](%)	97 (33/34)	86 (19/22)
Specificity[c](%)	89 (8/9)	100 (7/7)
Positive predictive value (PPV)[d](%)	97 (33/34)	100 (19/19)
Negative predictive value (NPV)[e](%)	89 (8/9)	70 (7/10)

Note: Criteria for congenital infection:
Termination of pregnancy: RV RNA and/or rubella virus detection in fetal tissue; live-born infant: CRS and specific IgM antibody detection at birth; no CRS-like symptoms and specific; IgM antibody detection at birth; and persistence of specific IgG antibodies to 7–11th months of age.
[a]RT-nPCR was performed with the protocol of Bosma et al. (1995a) and samples were tested in duplicate. FB was tested for rubella-specific IgM antibodies using three different commercial rubella IgM test kits, two based on the IgM-capture principle. IgM testing was done in serum dilution recommended by the manufacturer and in a lower dilution mainly in duplicate.
[b]Sensitivity: (number of true positives/total number of fetuses or newborns congenital infected) × 100.
[c]Specificity: (number of true negatives/total number of fetuses or newborns uninfected) × 100.
[d]PPV: (number of true positives/number of positive test results) × 100.
[e]NPV: (number of true negatives/number of negative test results) × 100.

per cent of those with positive RT-nPCR results were associated with fetal infection, while 89% of those with negative RT-nPCR results were uninfected. False-positive and false-negative RT-nPCR results have occurred in all studies (Table 2). Thus, a negative result in a single sample for PD, even if carried out under optimal conditions, can never exclude fetal infection.

Borderline results may be obtained when maternal infection has occurred between 12 and 18 weeks of gestation, particularly if infection was subclinical, as viral load and rubella IgM concentrations are more likely to be low in these situations (G. Enders, personal communication). In these cases repeat testing and collection of further samples are necessary, as demonstrated by Tang et al. (2003), who described negative results in amniotic fluid taken at 19 and 23 weeks gestation, but RV RNA and rubella IgM detected in fetal blood taken at 23 weeks. However, collection of further samples will delay diagnosis and must always be weighed against the risks incurred when carrying out invasive prenatal tests (Table 3).

Neonatal diagnosis and diagnosis in later infancy

Neonates born to women who have had rubella at any stage of pregnancy should be tested for evidence of congenital infection, as should neonates born with clinical features compatible with congenital rubella (Chapter 2). When these children are assessed age, clinical findings, maternal history, rubelliform illnesses since birth as well as laboratory results should be taken into consideration. Results of laboratory tests are required to confirm congenital rubella (Miller et al., 1994; World Health Organisation, 1999; Reef et al., 2000). Rubella IgM, persistence of specific IgG between 7 and 11 months of age, and detection of virus excretion by virus isolation or RT-nPCR are established methods for the diagnosis of congenital rubella (Table 5).

When there is a history of rubella or unconfirmed rubella-like illness in the mother, the neonate should be tested for rubella IgM using a sensitive IgM-capture assay, as this method appears to be more sensitive for the diagnosis of congenital rubella than indirect IgM EIAs (Chantler et al., 1982; Corcoran and Hardie, 2005). Rubella IgM may also be detected in oral fluid (Eckstein et al., 1996). Infants born with signs or symptoms suggestive of rubella-like defects should also be tested at the earliest opportunity, since there is a risk of RV transmission to susceptible persons.

Specific IgM is detected in most cases of CRS tested within the first 3 months after birth with all types of IgM tests (Table 6). If a low or equivocal IgM result is obtained, a further specimen should be obtained within a week and tested in additional IgM-capture tests. A negative IgM results in the neonatal period and up to 3 months of age will virtually exclude congenital infection. For samples collected between 3 and 18 months of age, IgM-capture EIAs are the most sensitive and specific methods available. Beyond 18 months rubella IgM is rarely detected. In *asymptomatic* congenital infection rubella IgM is, in our experience, detected up to 3 months of age at rates similar to those of CRS, but more rarely beyond 3 months of age.

Table 5

Laboratory techniques for the diagnosis of congenital rubella infection

Established methods	Prenatal diagnosis	Diagnosis in infancy	References
Rubella-specific IgM in serum	Yes	Birth–1 year	Chantler et al. (1982)[a] Cooper (1975)[a] Thomas et al. (1993)[a] Enders et al. (personal communication)[a]
Rubella-specific IgM in oral fluid	No	Birth–1 year	Eckstein et al. (1996)[a]
Rubella-specific IgG	No	7 months–1 year	
RV isolation in cell culture	Yes[b]	Birth–1 year	Cooper and Krugman (1967)[a]
RT-nPCR (RV RNA detection)	Yes	Birth–1 year	Enders et al. (personal communication)[a] Bosma et al. (1995b)[a, c] Tanemura et al. (1996)[c] Revello et al. (1997)[c] Katow (1998)[c] Macé et al. (2004)[c] Enders et al. (personal communication)[a,c]

[a] Diagnosis in infancy.
[b] Too slow to be useful.
[c] Prenatal diagnosis.

Laboratory confirmation of congenital rubella infection may also be obtained by isolation of RV or detection of RV RNA in nasopharyngeal swabs, urine and oral fluid. Although they were using relatively insensitive methods for virus isolation, Cooper and Krugman (1967) isolated RV from nasopharyngeal secretions of most cases of congenital rubella up to the age of 1 month, from 62% at 1–4 months, 33% at 5–8 months, 11% at 9–12 months and only 3% at 13–20 months of age. In more recent studies we have confirmed that RT-nPCR was more sensitive than virus isolation for all samples, with virus excretion detected more frequently and for longer in CRS cases (Table 7). The latest positive RT-nPCR result was obtained from a urine sample obtained at 23 months of age. In asymptomatic congenital rubella infection RV RNA was detected by RT-nPCR less frequently and only up to 3 months of age. RT-nPCR has also been successfully employed to detect RV RNA in postmortem fetal tissues (Bosma et al., 1995b; Revello et al., 1997). Specimens can be

Table 6

Frequency and persistence of rubella-specific IgM in congenitally infected infants with clinical features of CRS:data from four studies

Reference	Technique for detection of rubella IgM	No. sera positive/no. tested (%)					Total tested
		Age (months)[a]					
		<3	3–6	6–12	12–18	18–24	
Cooper (1975)	SDG+HAI	36/60 (60%)	10/23 (43.5%)	12/31 (38%)	5/21 (23%)	1/21 (5%)	143
Chantler et al. (1982)	MACRIA	35/35 (100%)	6/7 (85.7%)	13/21 (62%)	9/19 (47.4%)	1/17 (5.9%)	99
Thomas et al. (1993)	MACEIA	35/35 (100%)	4/7 (57%)	6/21 (28.6%)	2/19 (10.5%)	1/17 5.9%	126
	MACRIA	51/51 (100%)	11/13 (84.6%)	7/23 (30.4%)	4/15 (26.7%)	1/24 (4%)	
Enders et al. (unpublished data, 1985–2004)	Commercial IgM-capture and indirect IgM EIAs	91/95 (96%)	13/15 (87%)	10/14 (71%)	3/4 (75%)	4/6 (67%)	134

[a] Age ranges are approximate, as different age ranges were used in each study. SDG = sucrose density gradient centrifugation; HAI = haemagglutination inhibition; MACRIA = IgM-capture radioimmunoassay; and MACEIA = IgM-capture enzyme immunoassay.

Table 7

Rubella virus excretion detected by RT-nPCR and virus isolation in 22 cases of CRS (Enders et al., unpublished results, 1993–1996)

RT-nPCR	Months after birth									
	0–3		4–6		7–12		13–24		Total	
	%	(pos/n)	%	(pos/n)	%	(pos/n)	%	(pos/n)	%	(pos/n)
Sample										
Nasopharyngeal secretions	84	(16/19)	100	(2/2)	0	(0/1)	50	(3/6)	75	(21/28)
Urine	76	(19/25)	58	(7/12)	43	(3/7)	60	(3/5)	65	(32/49)
EDTA-blood	46	(6/13)	44	(4/9)	0	(0/2)	0	(0/1)	40	(10/25)
Cerebrospinal fluid	33	(2/6)	25	(1/4)	—		25	(1/4)	29	(4/14)
Virus isolation										
Nasopharyngeal secretions	37	(7/19)	0	(0/2)	0	(0/1)	0	(0/6)	25	(7/28)
Urine	36	(9/25)	33	(4/12)	14	(1/7)	40	(2/5)	33	(16/49)
EDTA-blood	25	(1/4)	33	(1/3)	—		0	(0/1)	25	(2/8)
Cerebrospinal fluid	25	(1/5)	50	(2/4)	—		25	(1/4)	31	(4/13)

pos/n = no. positive/no. tested.

transported from distant sites on dry ice or liquid nitrogen, and for RT-nPCR (but not virus isolation) they can be dried on filter paper (De Swart et al., 2001).

When cataracts are removed the lens aspirates can be used for detection of RV. RT-nPCR has been successfully employed to detect RV RNA in lens aspirates from infants aged 2–12 months with CRS (Bosma et al., 1995b). RV RNA detection has not been fully evaluated for detection of RV in older children with eye disease.

Prompt diagnosis of congenital rubella is more difficult when there is no history of a rubella-like illness in pregnancy, especially if the infant presents at more than 3 months of age, when it may no longer be possible to detect rubella IgM, isolate RV or detect RV RNA. This situation arises when children have isolated sensorineural deafness detected in later infancy. Such children should be tested for rubella IgG at 7–11 months of age, when maternal IgG antibody will normally have disappeared and MMR vaccination has not yet been received (Fig. 9); the detection of specific IgG is suggestive of congenital infection. The value of testing sera from older children will depend on the epidemiology of rubella locally. Although rubella is rare in children under the age of two in temperate climates, it is more frequent in tropical, developing countries. Congenital rubella infection may be excluded in most sero-negative infants and children although some congenital rubella patients will lose rubella antibody (see below). Diagnosis of congenital rubella in children who have received MMR vaccination is unlikely to be possible, although failure to develop rubella antibodies after vaccination is suggestive of congenital rubella (p. 68).

RV SP15 peptide EIA and rubella IB may also be useful for the diagnosis of congenital rubella and have been used to define serological differences between

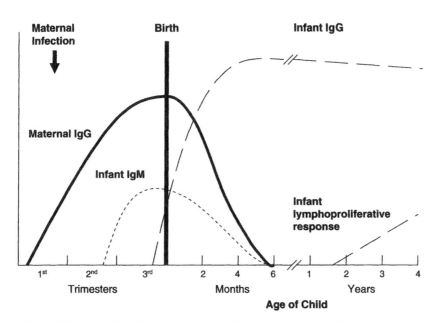

Fig. 9 Serological markers for diagnosis of congenital rubella. (Reproduced from Diagnostic Procedures for Viral, Rickettsial and Chlamydial Infections, 7th edition with permission from the American Public Health Association.)

infants with symptomatic and asymptomatic congenital rubella infection (Meitsch et al., 1997; Pustowoit and Liebert, 1998). In the 10 children with CRS born to mothers with infection in the first 12 weeks of gestation, antibodies to SP15 and the E2 glycoprotein were detected less frequently and in significantly lower concentrations than in the 8 asymptomatic congenital infected newborns of mothers with infection beyond the 10th week of gestation. However, these assays need further evaluation.

Detection of low-avidity IgG1 by the DEA-shift method between 6 months and 3 years of age may aid the diagnosis, as the avidity matures more slowly in children with CRS than following postnatal infection (Thomas et al., 1993). Tests for cell-mediated immunity may also be useful. In a study by O'Shea et al. (1992) 10 of the 13 (79.9%) children under the age of 3 with CRS did not respond in a rubella-specific lymphoproliferative assay, compared with only 1 of the 13 (7.6%) seropositive control children. Although this assay is more demanding than EIA, it may be useful for the diagnosis of congenital rubella in children under the age of 3. More recently, the enzyme-linked immunospot test (ELISPOT) has been used to detect rubella-specific cellular immunity (B. Pustowoit et al., personal communication, 2005). Unfortunately it is necessary to obtain heparinized blood for all cell functional assays and to perform the test within a very short time (24 h) after blood sampling.

Loss of rubella antibodies in congenital rubella.   The loss of rubella antibodies was first reported by Cooper et al. (1971), who showed that 50 of the 270 (18.5%)

congenitally infected children had lost HAI antibodies by the age of 5. When 19 of these seronegative children were given HPV77 rubella vaccine, 17 (89%) failed to seroconvert. Since then Ueda et al. (1987) reported that 13 of the 381 children with congenital rubella became seronegative by HAI between 3 and 17 years of age. Forrest et al. (2002) reported that 13 of 32 (41%) of Gregg's original congenitally infected patients were seronegative by EIA at the age of 60. Enders and colleagues have found that five hearing impaired pregnant women aged 21–34 years with congenital rubella infection and low levels of HAI, IgG and NT antibodies, failed to respond to first or repeated vaccination and in three of the five women tested by IB, the E2 band—a tentative marker for immunity—was absent (G. Enders, personal communication, 2003). Best and O'Shea (unpublished observations) also noted that adult women with a history of congenital rubella infection failed to seroconvert when vaccinated. Thus, the lack of rubella antibodies in older children may not exclude congenital rubella, but failure to respond to vaccine, may provide evidence of congenital rubella infection in a patient with rubella-like defects.

Remaining diagnostic problems

Protective antibodies

Further research is required to identify the type of specific antibody associated with protection from reinfection. This is of particular concern for populations with vaccine-induced immunity, which may wane in the absence of the booster effects due to contact with circulating RV. While previous natural rubella infection is generally considered to induce lifelong protection, this cannot be guaranteed following vaccination, although it is probable that immunological memory exists.

The reason why reinfection occurs in some individuals is not clear, since intranasal challenge studies have not identified any antibody type (NT, HAI, IgG, IgA) or titre, which correlates with protection against reinfection (Harcourt et al., 1980; O'Shea et al., 1983, 1994). Similarly, RV-specific lymphoproliferative responses do not appear to be associated with protection (O'Shea et al., 1994). Studies of the E1 open reading frame of RV strains from cases of reinfection suggest that these strains do not differ from other strains (Frey et al., 1998). The absence of antibodies against known linear epitopes or unknown conformational epitopes on the E1 and E2 glycoproteins are assumed to be associated with lack of protective immunity (Wolinsky et al., 1993; Gießauf et al., 2004). Studies of immune responses following rubella reinfection suggest that lack of E2 antibodies may be a marker of susceptibility to reinfection, however, more data are required to confirm this hypothesis.

Standardization

Standardization of most rubella methods is required. Commercial and "in-house" HAI tests differ by the use of different erythrocytes, types of antigen and method of absorption of patients' sera. Various rubella IgG EIAs are used worldwide for rubella antibody screening, but differences in calibration have affected the interpretation of results and have made the definition of an IgG cut-off level difficult (Pustowoit et al., 1996; Matter et al., 1997). While European kit manufacturers use the WHO second international standard for anti-rubella serum (Statens Seruminstitut, Copenhagen) or national substandards, until 1997 most US manufacturers calibrated their tests by the CDC standard (e.g. Gunter et al., 1996). As a result, some US tests achieved IgG concentrations that were twice as high as those of several European test kits (Pustowoit et al., 1996; G. Enders, unpublished observations). Thus, results from the same patient may have been reported as rubella antibodies "detected" in a US and "not detected" in a European rubella IgG EIA. This may also happen in patients with a IgG antibody value around cut-off level at repeat testing several months apart—even with the same test and in the same laboratory. In such cases the question of immunity may be decided on a history of at least one documented vaccination. In a study by the European Sero-Epidemiology Network involving seven laboratories, large differences in unitage were found even though all laboratories were using a standardized EIA method (Andrews et al., 2000), suggesting that other factors like antigen type and quality play a role in achieving uniform IgG results.

Diagnosis of congenital rubella

Detection of rubella IgM with IgM-capture techniques is usually a reliable technique for the diagnosis of congenital rubella in the first 3 months of life (see Neonatal diagnosis and diagnosis in later infancy). However, there is a demand for the evaluation of new techniques for diagnosis of congenital rubella after 3 months in children with deafness only, as they may not present in the neonatal period, as discussed above.

Conclusions and outlook

As a result of WHO recommendations (World Health Organisation, 2000), programmes for surveillance of rubella and congenital rubella have been established in many countries, with a continuing demand for techniques for the diagnosis of rubella and congenital rubella (World Health Organisation, 1999). Most problems encountered with rubella diagnosis can now be solved with the improved and expanded range of serological methods described in this chapter.

Although many countries have substantially reduced the incidence of rubella and congenital rubella, unsubstantiated fears about the safety of MMR vaccine in countries such as the UK, have led to a fall in uptake of MMR to below 80%. This is not high enough to prevent circulation of RV and may result in some unvaccinated girls remaining susceptible into their childbearing years. Recent rubella outbreaks have also been seen among two connected religious groups in the Netherlands and Canada, who refuse vaccination. Thus, it is important to maintain and improve techniques for the laboratory diagnosis of rubella and congenital rubella.

In addition, antenatal rubella antibody screening should be continued, even in countries with well-established vaccination programmes, as it is of particular importance for immigrant women from countries with endemic rubella and no rubella vaccination or poorly implemented vaccination programmes. Recent reports of symptomatic congenitally infected infants born to such women in European countries should draw attention to the fact that CRS is down but not out (Tookey et al., 2002; Tookey, 2004; G. Enders, personal observation, 2003–2005).

Acknowledgements

We wish to thank Simone Exler and Marion Biber of the Laboratory Professor G. Enders and Partners for their excellent assistance in preparing and correcting this chapter and Dr. Siobhan O'Shea, Guy's and St Thomas' NHS Foundation Trust for review of the manuscript.

References

Akingbade D, Cohen BJ, Brown DWG. Detection of low-avidity immunoglobulin G in oral fluid samples: new approach for rubella diagnosis and surveillance. Clin Diagn Lab Immunol 2003; 10: 189–190.

Andrews N, Pebody RG, Berbers G, et al. The European sero-epidemiology network: standardizing the enzyme immunoassay results for measles, mumps and rubella. Epidemiol Infect 2000; 125(1): 127–141.

Banatvala JE, Brown DWG. Rubella. Lancet 2004; 363: 1127–1137.

Ben Saleh A, Zaatour A, Pomery L. Validation of a modified commercial assay for the detection of rubella-specific IgG in oral fluid for use in population studies. J Virol Methods 2003; 114: 151–158.

Best JM, Banatvala JE. A comparison of RK13, vervet monkey kidney and patas monkey kidney cell cultures for the isolation of rubella virus. J Hyg (Camb) 1967; 65: 263–271.

Best JM, Banatvala JE. Serum IgM and IgG responses in postnatally acquired rubella. Lancet 1969; 1: 65–68.

Best JM, Banatvala JE, Morgan-Capner P, Miller E. Fetal infection after maternal reinfection with rubella: criteria for defining reinfection. Br Med J 1989; 299: 773–775.

Best JM, Castillo-Solorzano C, Spika JS, et al. Reducing the global burden of congenital rubella syndrome. Report of the WHO Steering Committee on Research Related to Measles and Rubella Vaccines and Vaccination, June 2004. J Infect Dis 2005; 192(11): 1890.

Best JM, O'Shea S. Rubella virus. In: Diagnostic Procedures for Viral, Rickettsial and Chlamydial Infections (Lennette EH, Lennette DA, Lennette ET, editors). 7th ed. Washington, DC: American Public Health Association 1995; pp. 583–600.

Best JM, O'Shea S, Tipples G, et al. Interpretation of rubella serology in pregnancy—pitfalls and problems. Br Med J 2002; 325: 47–48.

Bosma TJ, Best JM, Corbett KM, et al. Nucleotide sequence analysis of a major antigenic domain of 22 rubella virus isolates. J Gen Virol 1996; 77: 2523–2530.

Bosma TJ, Corbett KM, Eckstein MB, et al. Use of PCR for prenatal and postnatal diagnosis of congenital rubella. J Clin Microbiol 1995b; 33: 2881–2887.

Bosma TJ, Corbett KM, O'Shea S, et al. PCR for detection of rubella virus RNA in clinical samples. J Clin Microbiol 1995a; 33: 1075–1079.

Böttiger B, Jensen IP. Maturation of rubella IgG avidity over time after acute rubella infection. Clin Diagn Virol 1997; 8(2): 105–111.

Bullens D, Smets K, Vanhaesebrouck P. Congenital rubella syndrome after maternal reinfection. Clin Pediatr (Phila) 2000; 39(2): 113–116.

Chantler S, Evans CJ, Mortimer PP, Cradock-Watson JE, Ridehalgh MKS, Cooper LZ, Florman AL, Ziring PR, Krugman S. A comparison of antibody capture radio- and enzyme immunoassays with immunofluorescence for detecting IgM antibody in infants with congenital rubella. J Virol Methods 1982; 4: 305–313.

Cooper LZ. Congenital rubella in the United States. In: Progress in Clinical and Biological Research (Krugman S, Gerhson A, editors). 3. New York: Alan R Liss, Inc; 1975; pp. 1–21.

Cooper LZ, Forman AL, et al. Loss of rubella hemagglutination inhibition antibody in congenital rubella. Failure of seronegative children with congenital rubella to respond to HPV-77 rubella vaccine. Am J Dis Child 1971; 122: 397–403.

Cooper LZ, Krugman S. Clinical manifestations of postnatal and congenital rubella. Arch Ophthalmol 1967; 77: 434.

Cooray S, Warrener L, Jin L. Improved RT-PCR for diagnosis and epidemiological surveillance of rubella. J Clin Virol 2006; 35: 73–80.

Corcoran C, Hardie DR. Serologic diagnosis of congenital rubella: a cautionary tale. Pediatr Infect Dis J 2005; 24(3): 286–287.

Cordoba P, Lanoel A, Grutadauria S, Zapata M. Evaluation of antibodies against a rubella virus neutralizing domain for determination of immune status. Clin Diagn Lab Immunol 2000; 7(6): 964–966.

Crino JP. Ultrasound and fetal diagnosis of perinatal infection. Clin Obstet Gynecol 1999; 42: 71–80.

Cutts F, Brown DWG. The contribution of field tests to measles surveillance and control: a review of available methods. Rev Med Virol 1995; 5: 35–40.

de Oliveira SA, Siqueira MM, Brown DWG. Diagnosis of rubella infection by detecting specific immunoglobulin M antibodies in saliva samples: a clinic-based study in Niteroi RJ, Brazil. Rev Soc Bras Med Trop 2000; 33: 335–339.

De Swart RL, Nur Y, Abdallah A. Combination of reverse transcriptase PCR analysis and immunoglobulin M detection on filter paper blood samples allows diagnostic and epidemiological studies of measles. J Clin Microbiol 2001; 39(1): 270–273.

del Mar Mosquera M, de Ory F, Moreno M, Echevarria JE. Simultaneous detection of measles virus, rubella virus, and parvovirus B19 by using multiplex PCR. J Clin Microbiol 2002; 40: 111–116.

Department of Health Screening for Infectious Diseases in Pregnancy. Standards to Support the UK Antenatal Screening Programme. London: Stationery Office; 2003. Available from http://www.dh.gov.uk/assetRoot/04/09/20/49/04092049.pdf.

Eckstein MB, Brown DWG, Foster A, et al. Congenital rubella in south India: diagnosis using saliva from infants with cataract. Br Med J 1996; 312: 161.

Eggers M, Enders M, Strobel S, et al. A novel and rapid method for determination of rubella virus immunoglobulin G avidity. Clin Microbiol Infect 2005; 11(Suppl 2): 33.

Enders G. Problems of rubella diagnosis by various IgM techniques and the need for test combinations. In: Rapid Methods and Automation in Microbiology and Immunology (Habermehl KO, editor). Berlin: Springer; 1985; pp. 146–160.

Enders G. Vergleich eines einfachen Latex-Agglutinationstestes mit dem Röteln-Häagglutinationshemmtest und Hämolysis-in-Gel-Test. Lab Med 1987; 11: 221–223.

Enders G. Diagnostik von Rötelninfektionen in der Schwangerschaft durch konventionelle, immunologische und molekular-biologische Methoden. In: Neues in der Virusdiagnostik (Deinhardt F, Maass G, Spiess H, editors). Richard-Haas-Symposium der DVV; 1991; pp. 133–152.

Enders G. Qualitätssicherung in der Serodiagnostik bei der Mutterschaftsvorsorge: Qualitätssicherung und aktuelle Aspekte zur Serodiagnostik der Röteln in der Schwangerschaft. Symposium Moderne Aspekte der Mikrobiologischen Diagnostik, Kurzfassungen von Vorträgen des 3. Symposium am 04. Dezember 1996 in Berlin. Clin Lab 1997; 43: 1019–1032.

Enders, G. Mütterliche Infektionen mit dem Risiko der prä- und perinatalen Übertragung—Teil I. Labormedizinische Aspekte wichtiger Infektionen im Überblick. Gynäkologie + Geburtshilfe 06, 2005; 6: 38–46.

Enders G, Bäder U, Lindemann L, et al. Prenatal diagnosis of congenital cytomegalovirus infection in 189 pregnancies with known outcome. Prenat Diagn 2001; 5: 362–377.

Enders G, Jonatha W. Prenatal diagnosis of intrauterine rubella. Infection 1987; 15: 162–164.

Enders G, Knotek F. Detection of IgM antibodies against rubella virus: comparison of two indirect ELISAs and an anti-IgM capture immunoassay. J Med Virol 1986; 19: 377–386.

Enders G, Knotek F. Rubella IgG total antibody avidity and IgG subclass-specific antibody avidity assay and their role in the differentiation between primary rubella and rubella reinfection. Infection 1989; 17(4): 218–226.

Enders G, Knotek F, Pacher U. Comparison of various serological methods and diagnostic kits for the detection of acute, recent, and previous rubella infection, vaccination, and congenital infections. J Med Virol 1985; 16: 219–232.

Forrest JM, Turnbull FM, Sholler GF, et al. Gregg's congenital rubella patients 60 years later. Med J Aust 2002; 177: 664–667.

Frey TK, Abernathy ES, Bosma TJ, et al. Molecular analysis of rubella virus epidemiology across three continents, North America, Europe and Asia, 1961–1997. J Infect Dis 1998; 178: 642–650.

Gerna G, Zannino M, Revello M, et al. Development and evaluation of a capture enzyme-linked immunosorbent assay for determination of rubella immunoglobulin M using monoclonal antibodies. J Clin Microbiol 1987; 25: 1033–1038.

Gießauf A, Letschka T, Walder G, et al. A synthetic peptide ELISA for the screening of rubella virus neutralizing antibodies in order to ascertain immunity. J Immunol Methods 2004; 287: 1–11.

Gitmans U, Enders G, Glück H, et al. Rötelnschutzimpfung mit $HPV_{77}DE_5$- und RA 27/3-Impfstoffen von 11–16 Jahre alten Mädchen: Untersuchungen zur Antikörperentwicklung und Antikörperpersistenz nach 4 und 8 Jahren. Immun Infekt 1983; 11: 79–90.

Grangeot-Keros L, Enders G. Evaluation of a new enzyme immunoassay based on recombinant rubella virus-like particles for detection of immunoglobulin M antibodies to rubella virus. J Clin Microbiol 1997; 35(2): 398–401.

Grangeot-Keros L, Pustowoit B, Hobman T. Evaluation of Cobas Core Rubella IgG EIA recomb, a new enzyme immunoassay based on recombinant rubella-like particles. J Clin Microbiol 1995; 33: 2392–2394.

Gunter EW, Lewis BG, Koncikowski SM. Laboratory Procedures Used for the Third National Health and Nutrition Examination Survey (NHANES III), 1988–1994, U.S. department of health and human services (http://www.cdc.gov/nchs/data/nhanes/nhanes3/cdrom/nchs/manuals/labman.pdf); 1996.

Halonen PE, Ryan IH, Stewart JA. Rubella hemagglutinin prepared with alkaline extraction of virus grown in suspension culture of BHK-21 cells. Proc Soc Exp Biol Med 1967; 125: 162–167.

Harcourt G, Best JM, Banatvala JE. Rubella specific serum and nasopharyngeal antibodies in volunteers with naturally acquired and vaccine induced immunity following intranasal challenge. J Infect Dis 1980; 142: 145–155.

Hedman K, Rousseau SA. Measurement of avidity of specific IgG for verification of recent primary rubella. J Med Virol 1989; 27: 288–292.

Hedman K, Seppälä I. Recent rubella virus infection indicated by a low avidity of specific IgG. J Clin Immunol 1988; 8: 214–221.

Helfand RF, Keyserling HL, Williams I, Murray A, Mei J, Moscatiello C, Icenogle J, Bellini WJ. Comparative detection of measles and rubella IgM and IgG derived from filter paper blood and serum samples. J Med Virol 2001; 65(4): 751–757.

Henquell C, Bournazeau JA, Vanlieferinghen P, et al. Recrudescence en 1997 des infections rubéoliques pendant la grossesse. La Presse Méd 1999; 28(15): 777–780.

Hobman TC, Lundstrom ML, Mauracher CA, et al. Assembly of rubella virus structural proteins into virus-like particles in transfected cells. Virology 1994; 202(2): 574–585.

Hodgson J, Morgan-Capner P. Evaluation of a commercial antibody capture enzyme immunoassay for the detection of rubella specific IgM. J Clin Pathol 1984; 37: 573–577.

Hudson P, Morgan-Capner P. Evaluation of 15 commercial enzyme immunoassays for the detection of rubella-specific IgM. Clin Diagn Virol 1996; 5: 21–26.

Jacquemard F, Hohlfeld P, Mirlesse V, et al. Prenatal diagnosis of fetal infections. In: Infectious Diseases of the Fetus and Newborn Infant (Remington JS, Klein JO, editors). Philadelphia: WB Saunders Company; 1995; pp. 99–107.

Jauniaux E, Rodeck C. Use, risks and complications of amniocentesis and chorionic villous sampling for prenatal diagnosis in early pregnancy. Early Pregnancy 1995; 1(4): 245–252.

Jin L, Vyse A, Brown DW. The role of RT-PCR assay of oral fluid for diagnosis and surveillance of measles, mumps and rubella. Bull World Health Organ 2002; 80: 76–77.

Karapanagiotidis T, Riddell M, Kelly H. Detection of rubella immunoglobulin M from dried venous blood spots using a commercial enzyme immunoassay. Diagn Microbiol Infect Dis 2005; 53: 107–111.

Katow S. Rubella virus genome diagnosis during pregnancy and mechanism of congenital rubella. Intervirology 1998; 41: 163–169.

Kurtz JB, Mortimer PP, Mortimer PR, et al. Rubella antibody measured by radial haemolysis. Characteristics and performance of a simple screening method for use in diagnostic laboratories. J Hyg (Camb) 1980; 84(2): 213–222.

Macé M, Cointe D, Six C, et al. Diagnostic value of reverse transcription-PCR of amniotic fluid for prenatal diagnosis of congenital rubella infection in pregnant women with confirmed primary rubella infection. J Clin Microbiol 2004; 42(10): 4818–4820.

Matter L, Gorgievski-Hrisoho M, Germann D. Comparison of four enzyme immunoassays for detection of immunoglobulin M antibodies against rubella virus. J Clin Microbiol 1994; 32: 2134–2139.

Matter L, Kogelschatz K, Germann D. Serum levels of rubella virus antibodies indicating immunity: response to vaccination of subjects with low or undetectable antibody concentrations. J Infect Dis 1997; 175(4): 749–755.

McCarthy K, Taylor-Robinson CH, Pillinger SE. Isolation of rubella virus from cases in Britain. Lancet 1963; II: 593.

McKie A, Vyse A, Maple C. Novel methods for the detection of microbial antibodies in oral fluid. Lancet Inf Dis 2002; 2: 18–24.

Meitsch K, Enders G, Wolinsky JS, et al. The role of rubella-immunoblot and rubella-peptide-EIA for the diagnosis of the congenital rubella syndrome during the prenatal and newborn periods. J Med Virol 1997; 51(4): 280–283.

Meurman OH, Viljanen MK, Granfors K. Solid-phase radioimmunoassay of rubella virus immunoglobulin M antibodies: comparison with sucrose density gradient centrifugation test. J Clin Microbiol 1977; 5: 257–262.

Miller E, Cradock-Watson JE, Pollock TM. Consequences of confirmed maternal rubella at successive stages of pregnancy. Lancet 1982; 2(8302): 781–784.

Miller E, Tookey P, et al. Rubella surveillance to June 1994: third joint report from the PHLS and the National Congenital Rubella Surveillance Programme. Commun Dis Rep 1994; 4: R146–R152.

Morgan-Capner P, Crowcroft NS. PHLS Joint Working Party of the Advisory Committees of Virology and Vaccines and Immunisation, Guidelines on the management of, and exposure to, rash illness in pregnancy (including consideration of relevant antibody screening programmes in pregnancy). Commun Dis Public Health 2002; 5(1): 59–71. [http://www.hpa.org.uk/cdph/issues/CDPHvol5/No1/rash illness guidelines.pdf].

Morgan-Capner P, Miller E, Vurdien JE, Ramsay ME. Outcome of pregnancy after maternal reinfection with rubella. Commun Dis Rep 1991; 1(6): R57–R59.

Morris M, Cohen B, Andrews N, Brown D. Stability of total and rubella-specific IgG in oral fluid samples: the effect of time and temperature. J Immunol Methods 2002; 266: 111–116.

Mortimer PP, Parry JV. Non-invasive virological diagnosis: are saliva and urine specimens adequate substitutes for blood? Rev Med Virol 1991; 1: 73–78.

Nedeljkovic J, Jovanovic T, Mladjenovic S, et al. Immunoblot analysis of natural and vaccine-induced IgG responses to rubella virus proteins expressed in insect cells. J Clin Virol 1999; 14(2): 119–131.

Nokes DJ, Nigatu W, Abebe A, et al. A comparison of oral fluid and serum for the detection of rubella-specific antibodies in a community study in Addis Ababa, Ethiopia. Trop Med Internat Health 1998; 3: 258–267.

O'Shea S, Best JM, Banatvala JE. Viremia, virus excretion, and antibody responses after challenge in volunteers with low levels of antibody to rubella virus. J Infect Dis 1983; 148: 639–647.

O'Shea S, Best JM, Banatvala JE. A lymphocyte transformation assay for the diagnosis of congenital rubella. J Virol Methods 1992; 37: 139–148.

O'Shea S, Corbett KM, Barrow SM, et al. Rubella reinfection; role of neutralising antibodies and cell-mediated immunity. Clin Diagn Virol 1994; 2: 349–358.

O'Shea S, Dunn H, Palmer S, et al. Automated rubella antibody screening: a cautionary tale. J Med Microbiol 1999; 48: 1047.

Parker SP, Cubitt WD. The use of the dried blood spot sample in epidemiological studies. J Clin Pathol 1999; 52: 633–639.

Parkman PD, Buescher EL, Artenstein MS. Recovery of the rubella virus from recruits. Proc Soc Exp Biol Med 1962; 111: 225–230.

Parkman PD, Mundon FK, McCown JM, Buescher EL. Studies of rubella. II. Neutralization of the virus. J Immunol 1964; 93: 608–617.

Perry KR, Brown WG, et al. Detection of measles, mumps and rubella antibodies in saliva using antibody capture radioimmunoassay. J Med Virol 1993; 40: 235–240.

Pitcovski J, Shmueli E, Krispel S, Levi N. Storage of viruses on filter paper for genetic analysis. J Virol Methods 1999; 83: 21–26.

Pustowoit B, Grangeot-Keros L, Hobman TC, Hofmann J. Evaluation of recombinant rubella-like particles in a commercial immunoassay for the detection of anti-rubella IgG. Clin Diagn Virol 1996; 5: 13–20.

Pustowoit B, Liebert UG. Predictive value of serological tests in rubella virus infection during pregnancy. Intervirology 1998; 41(4, 5): 170–177.

Ramsay M, Reacher M, O'Flynn C, et al. Causes of morbilliform rash in a highly immunised English population. Arch Dis Child 2002; 87: 202–206.

Ramsay ME, Brugha R, Brown DWG, et al. Salivary diagnosis of rubella: a study of notified cases in the United Kingdom, 1991–4. Epidemiol Infect 1998; 120: 315–319.

Reef SE, Plotkin S, Cordero JF, et al. Preparing for elimination of congenital rubella syndrome (CRS): summary of a workshop on CRS elimination in the United States. Clin Infect Dis 2000; 31: 85–95.

Revello MG, Baldanti F, Sarasini A, et al. Prenatal diagnosis of rubella virus infection by direct detection and semiquantitation of viral RNA in clinical samples by reverse transcription-PCR. J Clin Microbiol 1997; 35: 708–713.

Robert-Koch-Institut. Bekämpfung der Masern und Konnatalen Röteln: WHO-Strategie in der europäischen Region und aktueller Stand in Deutschland. Epidem Bull 2004; 10: 79–84.

Robertson SE, Featherstone DA, Gacic-Dobo M, Hersh BS. Rubella and congenital rubella syndrome: global update. Pan Am J Public Health 2003; 14(5): 306–315.

Schmidt M, Lidqvist C, Salmi A, Oker-Blom C. Detection of rubella virus-specific immunoglobulin M antibodies with a baculovirus-expressed E1 protein. Clin Diagn Lab Immunol 1996; 3(2): 216–218.

Skendzel LP. Rubella immunity. Defining the level of protective antibody. Am J Clin Pathol 1996; 106(2): 170–174.

Starkey WG, Newcombe J, Corbett KM, et al. Use of rubella E1 fusion proteins for the detection of rubella antibodies. J Clin Microbiol 1995; 33: 270–274.

Stewart GL, Parkman PD, Hopps HE, et al. Rubella-virus hemagglutination-inhibition test. N Engl J Med 1967; 276: 554–557.

Tanemura M, Suzumori K, Yagami Y, Katow S. Diagnosis of fetal rubella infection with reverse transcription and nested polymerase chain reaction: a study of 34 cases diagnosed in fetus. Am J Obstet Gynecol 1996; 174: 578–582.

Tang JW, Aarons E, Hesketh LM, et al. Prenatal diagnosis of congenital rubella infection in the second trimester of pregnancy. Prenat Diagn 2003; 23: 509–512.

Tedder RS, Yao JL, Anderson MJ. The production of monoclonal antibodies to rubella haemagglutinin and their use in antibody-capture assays for rubella-specific IgM. J Hyg (Camb) 1982; 88(2): 335–350.

Thomas HIJ, Morgan-Capner P. Rubella-specific IgG subclass avidity ELISA and its role in the differentiation between primary rubella and rubella reinfection. Epidemiol Infect 1988; 101: 591–598.

Thomas HIJ, Morgan-Capner P. Rubella-specific IgG1 avidity: a comparison of methods. J Virol Methods 1991; 31: 219–228.

Thomas HIJ, Morgan-Capner P, Cradock-Watson JE, et al. Slow maturation of IgG1, avidity and persistence of specific IgM in congenital rubella: implications for diagnosis and immunopathology. J Med Virol 1993; 41: 196–200.

Thomas HIJ, Morgan-Capner P, Enders G, et al. Persistence of specific IgM and low avidity specific IgG1 following primary rubella. J Virol Methods 1992a; 39: 149–155.

Thomas HIJ, Morgan-Capner P, Roberts A, Hesketh L. Persistent rubella-specific IgM reactivity in the absence of recent primary rubella and rubella reinfection. J Med Virol 1992b; 36: 188–192.

Tipples GA, Hamkar R, Mohktari-Azad T, et al. Evaluation of rubella IgM enzyme immunoassays. J Clin Virol 2004; 30: 233–238.

Tischer A, Gerike E. Immune response after primary and re-vaccination with different combined vaccines against measles, mumps, rubella. Vaccine 2000; 18: 1382–1392.

Tookey PA, Cortina-Borja M, Peckham CS. Rubella susceptibility among pregnant women in North London. J Public Health Med 2002; 24(3): 211–216.

Tookey P. Rubella in England, Scotland and Wales. Eurosurveillance Monthly 2004; 9(4): 21–22.

Tzeng WP, Zhou Y, Icenogle J, Frey TK. Novel replicon-based reporter gene assay for detection of rubella virus in clinical specimens. J Clin Microbiol 2005; 43(2): 879–885.

Ueda K, Tokugawa K, Fukushige J, et al. Hemaglutination inhibition antibodies in congenital rubella syndrome. Am J Dis Child 1987; 141: 211–212.

Vaananen P, Haiva VM, Koskela P, Meurman O. Comparison of a simple latex agglutination test with hemolysis in-gel, hemagglutination inhibition, and radioimmunoassay for detection of rubella virus antibodies. J Clin Microbiol 1985; 21(5): 793–795.

Vejtorp M, Fanoe E, Leerhoy J. Diagnosis of postnatal rubella by the enzyme-linked immunosorbent assay for rubella IgM and IgG antibodies. Acta Pathol Microbiol 1979; 87: 155–160.

Vesikari T, Vaheri A. Rubella: a method for rapid diagnosis of a recent infection by demonstration of the IgM antibodies. Br Med J 1968; 1: 221.

Vyse AJ, Brown DWG, Cohen BJ, et al. Detection of rubella virus-specific immunoglobulin G in saliva by an amplification-based enzyme-linked immunosorbent assay using monoclonal antibody to fluorescein isothiocyanate. J Clin Microbiol 1999; 37: 391–395.

Vyse AJ, Cohen BJ, Ramsay ME. A comparison of oral fluid collection devices for use in the surveillance of virus diseases in children. Public Health 2001; 115: 201–207.

Vyse AJ, Jin L. An RT-PCR assay using oral fluid samples to detect rubella virus genome for epidemiological surveillance. Mol Cell Probes 2002; 16: 93–97.

Weller TH, Neva FA. Propagation in tissue culture of cytopathic agents form patients with rubella-like illness. Proc Soc Exp Biol Med 1962; 111: 215–225.

Wolinsky JS, Sukholutsky E, Moore WT, et al. An antibody- and synthetic peptide-defined rubella virus E1 glycoprotein neutralization domain. J Virol 1993; 67(2): 961–968.

World Health Organisation. Guidelines for Surveillance of Congenital Rubella Syndrome and Rubella. Field test version. WHO/V&B/99.22. Geneva: WHO; 1999.

World Health Organisation. Preventing congenital rubella syndrome. Wkly Epidemiol Rec 2000; 75: 290–295.

World Health Organisation. Standardization of the nomenclature for genetic characteristics of wild-type rubella viruses. Wkly Epidemiol Rec 2005; 80: 125–132.

Zhang T, Mauracher CA, Mitchell LA, Tingle AJ. Detection of rubella virus-specific immunoglobulin G (IgG), IgM, and IgA antibodies by immunoblot assays. J Clin Microbiol 1992; 30(4): 824–830.

Zheng DP, Frey TK, Icenogle J, et al. Global distribution of rubella virus genotypes. Emerg Infect Dis 2003; 9(12): 1523–1530.

Zrein M, Joncas JH, Pedneault L, et al. Comparison of a whole-virus enzyme immunoassay (EIA) with a peptide-based EIA for detecting rubella virus immunoglobulin G antibodies following rubella vaccination. J Clin Microbiol 1993; 31(6): 1521–1524.

Rubella Viruses
Jangu Banatvala and Catherine Peckham (Editors)
© 2007 Elsevier B.V. All rights reserved
DOI 10.1016/S0168-7069(06)15004-1

Chapter 4

Rubella Vaccine

Susan Reef[a], Stanley A. Plotkin[b]

[a] *Rubella and Mumps Activity, National Immunization Program, Centers for Disease Control and Prevention, 1600 Clifton Road, Atlanta, GA 30333, USA*
[b] *University of Pennsylvania and Sanofi Pasteur, 4650 Wismer Road, Doylestown, PA 18901, USA*

Introduction

Rubella was initially thought to be one of the most benign rash illnesses. However, in 1941, Gregg associated congenital cataracts with maternal rubella, discovering that rubella virus was one of the most teratogenic agents known (Gregg, 1941). Twenty years later, in 1962, the rubella virus was isolated by two independent groups, Weller and Neva at Harvard University School of Public Health and Parkman, Buescher and Arenstein at Walter Reed Army Institute for Research (Parkman et al., 1962; Weller and Neva, 1962). This achievement paved the way for the development of serological tests and vaccines. Meanwhile, a worldwide rubella pandemic, which started in Europe in 1962–1963, spread to the U.S. in 1964 and 1965. The U.S. epidemic resulted in approximately 12.5 million rubella cases, 11,000 fetal deaths and 20,000 congenital rubella syndrome (CRS) cases (Rubella surveillance, 1969). This pandemic resulted in the accumulation of extensive knowledge on rubella and CRS and reinforced the urgency of developing a rubella vaccine.

Active immunization

To develop a vaccine, researchers approached the issue in two different ways: inactivated or killed vaccine or live-attenuated virus vaccine.

Inactivated virus vaccine

Shortly after the isolation of the virus, investigators attempted to develop an in-activated virus vaccine, but their attempts were thwarted (Sever et al., 1963; Meyer et al., 1969). Either the vaccines were not antigenic or if antibodies were produced, it was questioned if the preparation was contaminated with live virus. Not until 1980s–1990s did researchers understand the molecular and antigenic characteristics of the rubella virus, which might serve as a basis for development of alternative vaccines (Waxman and Wolinsky, 1985). Recently, two rubella virus DNAs were constructed and evaluated in a mouse model. Equivalent antibody responses in mice infected with natural rubella virus and DNA vaccines were documented with persistence of titers for 7 months (Pougatcheva et al., 1999). If developed these non-replicating rubella virus vaccines might be indicated for special circumstances, such as vaccination of pregnant women.

Live-attenuated virus vaccine

Development

After the isolation of the rubella virus, several groups set out to develop a live-attenuated vaccine. In 1965, Parkman et al. were the first to successfully attenuate the rubella virus with viral passage 77 times in African green monkey kidney cultures (Parkman et al., 1966). In 1969–1970, three vaccines were licensed in the U.S., including HPV-77 (dog kidney), HPV-77 (duck embryo) and Cendehill (rab-bit kidney) (Preblud et al., 1980). Shortly thereafter, the RA 27/3 vaccine (human diploid cells) was licensed in Europe (Plotkin et al., 1969). In Japan, the initial vaccines licensed were the Takahashi (rabbit kidney) and Matsuura (Japanese quail-embryo fibroblasts) vaccines (Perkins, 1985).

   Within 10 years after the initiation of the rubella vaccination program in the U.S., all three vaccines were replaced by RA 27/3, which had been licensed in Europe from 1970 onwards. In the U.S., the HPV-77 DK vaccine was withdrawn due to higher incidence of side effects as compared to the other vaccines. These side effects included arthritis, arthralgia and neuropathy. Shortly thereafter, the Cende-hill strain vaccine was withdrawn. In January 1979, RA 27/3 vaccine replaced HPV-77 DE vaccine and is currently the only licensed rubella vaccine in the U.S. (Preblud et al., 1980) and in most of the world.

   The virus that served as the basis for RA 27/3 vaccine was isolated from the 27th fetus studied by S. Plotkin during the 1964 rubella epidemic (Plotkin et al., 1969). The virus was cultured from the kidney tissue that was the third fetal ex-plant. The supernatant fluid from the kidney cells was directly planted on WI-38 human diploid cells. Attenuation was accomplished by serial passage at low tem-peratures (cold adaptation). The vaccine virus was produced between the 25th and 33rd passage in human diploid cells.

After the development and licensure of the initial rubella vaccines globally, additional vaccines have been licensed in various geographic locations. In 1980, a rubella vaccine (BRD-2) was developed in China using a rubella virus strain isolated from a child. The rubella virus strain was isolated on human diploid cells. In a trial comparing the BRD-2 vaccine and RA 27/3 vaccine, the seroconversion rate and mild side effects were similar (Yaru et al., 1985). In Japan, currently five different rubella vaccines are in use including TO-366 virus vaccine (Kakizawa et al., 2001).

Dose and route of administration

Worldwide, licensed rubella vaccines are administered subcutaneously with at least 1000 plaque form units (PFU) at the time of delivery. During the initial clinical trials, subcutaneous and intranasal routes of administration were evaluated. RA 27/3 vaccine was the only vaccine that produced secretory IgA, which may have a role in preventing reinfection (Hillary, 1971; Puschak et al., 1971).

Because of accelerated measles control efforts globally, various non-percutaneous routes (e.g., intranasal, aerosol, conjunctival) of administration are being evaluated (Cutts, 1997). A study by Dilraj documented that Edmonston–Zagreb (EZ) vaccine by aerosol produced better short term and persistence of antibody titer when compared to EZ or Schwartz measles vaccine subcutaneously (Dilraj et al., 2000). Because of the interest of combining rubella vaccine with measles vaccine, Mexican researchers have conducted studies aerosolizing both measles and rubella vaccine and have achieved the same seroconversion rates as after subcutaneous injection (Sepulveda J, Valdespino JL, personal communication, 2001). However, one of the concerns about aerosolizing rubella-containing vaccine is the risk of exposing and infecting susceptible pregnant women.

Combination of the vaccine

Rubella-containing vaccines are available as a single-antigen rubella-only or in combination as either measles–mumps–rubella (MMR) or measles–rubella (MR) or mumps–rubella (MR). In the U.S. and increasingly elsewhere, the trivalent vaccine (MMR) is recommended for use in the pediatric schedule. Additionally, as part of many of the adult or childhood accelerated measles campaign, rubella vaccine has been included as either MR or MMR vaccine. The MMR vaccine formulation (MMR II, Merck, Sharp and Dohme) used in the U.S. contains the Moraten-attenuated measles virus (1000 $TCID_{50}$), the Jeryl Lynn mumps vaccine (5000 $TCID_{50}$) and the RA 27/3 rubella virus (1000 $TCID_{50}$). A measles–rubella combined vaccine (M-R-Vax II, Merck, Sharpe and Dohme) as well as measles–mumps–rubella–varicella vaccine are available.

Outside the United States, several formulations are available. In Europe and elsewhere, available formulations include Trimovax (Sanofi-Pasteur), which contains the Schwarz measles virus (1000 $TCID_{50}$), the Urabe mumps virus (20,000 $TCID_{50}$) and the RA 27/3 rubella vaccine virus (1000 $TCID_{50}$); and Priorix (GlaxoSmithKline), which contains Schwartz measles virus (1000 $TCID_{50}$), the RIT 4385 (derived from Jeryl Lynn, 10,000 $TCID_{50}$) and the RA 27/3 rubella vaccine virus (1000 $TCID_{50}$). Berna also supplies RA 27/3 rubella vaccine.

Another major manufacturer is the Serum Institute of India, which manufactures three different rubella-containing formulations: rubella-only vaccine (Wistar RA 27/3, 1000 $CCID_{50}$), MR (for export only) (L-Zagreb, 5000 $CCID_{50}$) and MMR (Trestivac, EZ measles virus, 1000 $CCID_{50}$).

Immune responses

Immune reactions produced by vaccination include both IgM and IgG antibody classes and cellular immune responses. Several different assays have been used to measure the immunologic response including hemagglutination assay, enzyme-linked immunoabsorbent assay (EIA) and neutralization techniques.

The hemagglutination inhibition and neutralizing responses technique were used to measure antibody response for the earlier vaccines (HPV-77, Cendehill, RA 27/3); however, as more sensitive and easier to use tests became available, they were replaced with tests such as the EIA assay and radial diffusion. In many studies, 95–100% of persons aged 11 months and older who were vaccinated with RA 27/3 seroconverted (Weibel et al., 1980).

In countries that use MMR vaccine for their pediatric schedule, it is recommended that infants be vaccinated at 12 months or older to avoid interference by passively acquired maternal antibodies. However, in many countries a majority of the mothers are now vaccinated and there is quicker loss of antibody titers in infants of those mothers than in infants whose mothers were infected with wild rubella virus. Because of this situation, several investigators have evaluated the seroconversion rates of MMR vaccine in children < 12 months of age (Singh et al., 1994; Forleo-Neto et al., 1997). These studies show that the post-vaccination seroconversion rate is not significantly different in children aged 12 months as compared to children aged 15 months. However, in the Forleo-Neto study, infants aged 9 months had lower post-vaccination titers to rubella than children aged 15 months.

Several studies have compared antibody responses to different vaccines, particularly to those which were used earlier such as HPV-77 and Cendehill (reviewed by Plotkin, 2004). Recent studies have been conducted comparing vaccines incorporating rubella, mumps and measles. Priorix vaccine (GlaxoSmithKline) incorporates the RA27/3 strain of rubella, the Schwarz strain of measles and the RIT 4385 strain of mumps (related to Jeryl Lynn). Triviraten (Berna) incorporates RA27/3, the EZ (measles) and the Rubini (mumps). When compared with MMR II, this and the Priorix vaccine had comparable antibody responses. Many antibody assays are useful to measure titer levels; however, neutralizing antibodies may have biological significance by measuring functionality of the antibody. In comparing different vaccines, RA 27/3 vaccine induces higher neutralizing antibodies, but at a lower level than natural disease (Horstmann et al., 1985).

Rubella-specific IgM antibodies can be detected within 14 days after vaccination, peaking at 1 month, with variation dependent on the technique used (Enders, 1985). IgM antibodies are usually gone by 3 months post-vaccination. However, using an M-antibody capture radioimmunoassay (MACRIA), some investigators have been able to detect low levels of rubella IgM antibodies in 18 of 53 (34%) of vaccinees 1 year after vaccination and in 7 of 18 (39%) of vaccinees 3 years after vaccination (Best, 1991). Rubella IgM antibodies can be detected in both primary and reinfection; however, in reinfections, the IgM responses is transient and titers are generally lower (Balfour et al., 1981; O'Shea et al., 1994).

Viremia after vaccination occurs within 7–11 days after inoculation, but at a low level. However, pharyngeal excretion of virus is more frequent and may occur between 7 and 21 days. Because of the concern that the vaccine virus would be transmissible to susceptible contacts, several studies to evaluate contagiousness were conducted in family settings (Plotkin et al., 1969; Veronelli, 1970) and institutions. There was no evidence of spread of vaccine to the susceptible contacts. In one immunization campaign with almost 100,000 children vaccinated, only one of the 121 seronegative exposed pregnant women seroconverted (Scott and Byrne, 1971). She was among the 31 pregnant women who had close contact with a vaccinee. However, her elevated titer was documented only 14 days after exposure to her son who had been vaccinated, which is too soon to implicate transmission of the vaccine virus.

Few studies have been conducted documenting the cellular immune responses in persons who have been vaccinated or who have had natural infection (Lalla et al., 1973; Honeyman et al., 1974; Buimovici-Klein and Cooper, 1985). Those studies showed that natural infection and HPV-77, Cendehill and RA 27/3 vaccines produce cell-mediated responses with RA 27/3 being the closest to natural infection.

Protective effects and reinfection

The protective efficacy of rubella vaccination has been evaluated during outbreaks and by intranasal challenges of vaccinated volunteers with attenuated and unattenuated viruses.

The efficacy of rubella-containing vaccine has been estimated to approximately 95% in outbreak settings. Additionally, when rubella vaccination was initiated during the outbreak, rubella incidence dropped 2–3 weeks later among the vaccinated group (Furukawa et al., 1970; Landrigan et al., 1974). In both naturally infected and vaccinated individuals, reinfection occurs. However, the significance of reinfections is hotly debated. During an institutional outbreak of rubella, 5 of the 22 subjects (23%) previously vaccinated with HPV-77 or Benoit strains and one of the 66 (1%) naturally immune subjects, had subclinical reinfections without viremia (Davies et al., 1971).

Persons with natural infection are less likely to have reinfections than vaccinees and among the vaccine strains, RA 27/3 vaccine permits the fewest reinfections

(5%). In earlier studies of HPV-77 and Cendehill vaccines, the reinfection rate of vaccinated persons was 50% or greater (Fogel et al., 1978).

Reinfection is mainly a concern when women who were either previously vaccinated or post-natural infection have had maternal reinfection during pregnancy. In some of these situations congenital rubella was the result (Eilard and Strannegard, 1974; Forsgren et al., 1979; Bott and Eisenberg, 1982; Enders et al., 1984; Gilbert and Kudesia, 1989; Das et al., 1990; Keith, 1991; Morgan-Capner et al., 1991; Weber et al., 1993; Braun et al., 1994; Paludetto et al., 1994). However, many more women who were reinfected during pregnancy have delivered infants without evidence of congenital infection.

To fully understand the risk factors for reinfection, several challenge studies have been conducted. O'Shea and colleagues measured rubella immunity in reinfected women and seropositive volunteers who underwent a challenge with rubella virus. The presence or absence of neutralizing antibodies or cell-mediated response did not correlate with reinfection (O'Shea et al., 1994). Other researchers have noted that the risk of reinfection is correlated with lack of prior antibodies to a peptide of the E1 protein (Mitchell et al., 1996). In summary, reinfection does occur and infants with CRS have resulted from maternal reinfection, but these cases are rare compared to the number of women vaccinated who are protected.

Adverse events

The most common adverse events after vaccination include low-grade fever, rash, arthropathy and lymphadenopathy. Of all the adverse events, the occurrence of acute arthropathy and chronic arthritis are the most concerning, particularly in post-pubertal females.

In 1991, the Institute of Medicine reviewed data on four possible adverse events associated with rubella vaccine. These included acute arthritis, chronic arthritis, neuropathies and thrombocytopenia. The committee concluded that RA 27/3 vaccine causes acute arthritis. With regard to chronic arthritis, the committee concluded that, "Evidence is consistent with a causal relation between the currently used rubella vaccine strain (RA27/3) and chronic arthritis in adult women, although the evidence is limited in scope and confined to reports from one institution." The committee recommended that a double-blinded study be conducted to evaluate the association of chronic arthropathy associated with rubella vaccine (Howson et al., 1991).

Arthropathy

Acute arthropathy including arthritis and arthralgias may occur in up to 70% of persons with natural rubella, particularly post-pubertal females. Because of the frequency of joint manifestations in natural rubella infection, initial clinical trials evaluated and documented the frequency of acute arthropathy in those trials. The frequency of joint symptoms in adult females increases with age and is dependent on which vaccine is administered. HPV-77 DK vaccine had the greatest frequency of joint symptoms with 49% of susceptible vaccinees affected, followed by HPV-77-DE (30%), RA 27/3 (14%) and the lowest in Cendehill (9%). In either natural- and vaccine-associated disease, these symptoms usually have not caused disruption of activities and most often have not persisted. An extensive review of acute arthropathy and arthritis associated with rubella vaccine has been previously published (Howson et al., 1991).

During the mid-1980s, investigators from one institution reported persistent or chronic arthropathy in 5–11% of adult females following rubella vaccination (Tingle et al., 1986; Mitchell et al., 1993). To evaluate this finding, several researchers conducted studies. In a double-blinded case-control study, Slater and colleagues compared a group of approximately 500 women who had received rubella vaccine post-partum with a control group of women who were immune and did not receive rubella vaccine post-partum (Slater et al., 1995). Arthritis was defined as "a report of pain in any joint accompanied by at least of the following: limitation of motion, stiffness, local redness, heat or swelling: and which caused limitation of activity or led to a professional consultation." In the vaccine group, 19 (3.9%) and 16 (3.2%) in the control group met the case definition for arthritis. The difference between the groups for arthropathy was not statistically significant.

Ray et al. conducted a retrospective cohort study comparing three different groups of women: seronegative immunized, seronegative unimmunized and seropositive unimmunized (Ray et al., 1997). They found no evidence of a causal relationship between immunization with the RA 27/3 strain of rubella virus and persistent joint symptoms. Tingle et al. conducted a randomized double-blind, placebo-controlled study enrolling 636 women (Tingle et al., 1996). Of these women, 543 completed the one-month follow-up and 456 completed the 12-month follow-up. The frequency of acute arthropathy was 20% of the women in the placebo group and 30% of the women in the vaccinated group, which was statistically significant. "Persistent arthropathy was defined as occurrence of arthralgia or arthritis at any time during the 12 month after vaccination in women who experienced acute arthropathy and for whom joint complaints could not be attributed to other causes." The frequency of chronic arthropathy in the placebo group was 15% and 22% in the vaccine arm. However, whereas of 81 women in the vaccine group who developed acute arthropathy, 58 (72%) developed chronic arthropathy, 41 (75%) of the 55 women in the placebo arm who developed acute arthropathy also developed chronic arthropathy. In addition, there was no description of the severity or frequency or duration of the symptoms of the women in either group. By the case definition, a woman who had an episode of acute arthralgias for 3 days and then one episode of arthralgias of any duration after the one-month follow-up met the chronic or persistent arthropathy case definition. Thus, it is unclear what relationship these findings have to the chronic arthropathy previously described. To summarize, data from studies in the United States and experience from other countries using the RA 27/3 strain rubella vaccine have not supported an association, suggesting that chronic arthropathy may not be related to administration of rubella-containing vaccines.

Contraindications and precautions

General

Persons that have had an anaphylactic reaction following a dose of rubella-containing vaccine or to a vaccine component (e.g., gelatin, neomycin) should not receive rubella-containing vaccines such as single-antigen rubella, MR or MMR (Kelso et al., 1993; Sakaguchi et al., 1995). In persons who have an egg allergy, the risk of serious allergic reactions from measles- or mumps-containing vaccines is extremely low (Greenberg and Birx, 1988; Lavi et al., 1990); however, single-antigen rubella vaccine is made in human fibroblasts cells and there is no risk for persons with egg allergies (Plotkin, 2004).

Because of the possible risk of enhanced vaccine viral replication, persons with immunodeficiency diseases or persons who are immunosuppressed should not be vaccinated with MMR. However, certain persons in these groups (HIV-infected, steroid treated and leukemics) may be vaccinated under certain circumstances. In asymptomatic persons with HIV, who are severely immunocompromised and lack measles antibodies, MMR vaccine is recommended (CDC, 1998). Although the minimum dose or duration of treatment with steroids that causes immunosuppression is not well defined, treatment with low dose or alternate day therapy, or with topical or aerosol therapy, is not a contraindication to vaccination. Persons with leukemia may be vaccinated if they lack antibodies to measles, mumps or rubella and the time period between termination of therapy and vaccination is 3 months or longer. Because of concern about interference with seroconversion, rubella-containing vaccine should be given either 2 weeks before, or postponed for at least 3 months after administration of blood products (e.g., immunoglobulin, whole blood).

Pregnancy

Pregnancy is a contraindication to rubella vaccine. Because of the theoretical risk to the fetus, women are counseled to avoid pregnancy for 1 month after receipt of rubella-containing vaccine in the U.K., Canada and U.S. and 2 months in France. Because inadvertent vaccination of pregnant women does occur, many countries established registries to follow the outcome of pregnancies for women vaccinated 3 months prior to pregnancy or during pregnancy. In the U.S., between January 1971 and April 1989, the rubella vaccination in pregnancy registry recorded 321 susceptible pregnant women who delivered 324 infants. No defect associated with CRS was found. Of the 222 infants in whom serology was obtained, six (2.7%) had subclinical infection. All six were normal on physical examination and no defects were identified after 2 or more years of follow-up. These data are consistent with results reported from other countries, suggesting that if live-attenuated rubella vaccine causes defects associated with CRS, it does so at a very low rate ($< 1.3\%$) (CDC, 2001).

In 2001, the U.S. Advisory Committee on Immunization Practices recommended that the U.S. decrease the waiting period between vaccination and conception from 3 months to 1 month. This decision was based on data from the U.S. registry, unpublished data from the U.K. National Congenital Rubella Surveillance Programme (Tookey, 2004), Sweden and Germany (Enders et al., 1984). Data on 680 infants born to susceptible women who were inadvertently vaccinated 3 months prior to pregnancy or during pregnancy were examined. None of the infants was born with CRS. Limiting the analysis to the 293 infants born to susceptible mothers 1–2 weeks prior to conception and 4–6 weeks after conception, the maximum theoretical risk was 1.3%. This risk was substantially less than the > 50% risk for CRS associated with proven maternal infection during the first 20 weeks of pregnancy (CDC, 2001). Although there has been no observed risk of CRS in infants of mothers who were inadvertently vaccinated, Hofmann reported the first case of persistent fetal infection without defects following inadvertent vaccination in the 3 weeks after conception. The infant had been followed more than 14 months without any apparent defects or late manifestations of intrauterine rubella infection (Hofmann et al., 2000).

In view of the importance of protecting women of childbearing age from rubella, reasonable practices for avoiding vaccination of pregnant women in a rubella immunization program should include (1) asking women if they are pregnant, (2) excluding from the program those who say they are and (3) explaining the theoretical risks to the others before vaccinating. However, no evidence yet exists that RA 27/3 damages the fetus.

Use of rubella-containing vaccine globally

Since the licensure of the rubella vaccine in 1969, the global epidemiology of rubella and CRS has changed dramatically. Nevertheless, an estimated 100,000 cases of CRS occur each year throughout the world (Cutts et al., 1997). In 2000, WHO convened a meeting to review the worldwide status of CRS and its prevention (WHO, 2000). This meeting was prompted by the availability of more data on the CRS disease burden in developing countries, an increase in the number of countries with national rubella immunization programs and advances in laboratory diagnosis since the last international meeting on CRS and rubella in 1984. In 1996, 78 (36%) countries/territories used rubella vaccine in their national immunization programs. As of December 2003, 111 (57%) of member countries and territories used rubella vaccine. The proportion of countries using rubella vaccine varies markedly by WHO region: Africa—4%; South East Asia—18%; Eastern Mediterranean—57%; Western Pacific—52%; European—90%; and Americas—97%. The proportion of the world's population living in countries with national rubella vaccination programs has doubled from 12% in 1996 to 26% in 2004 (Fig. 1).

Much of the recent increase has occurred in Central and South America, in Europe and in the Western Pacific Region where polio eradication has allowed national immunization programs to address new challenges.

S. Reef, S.A. Plotkin

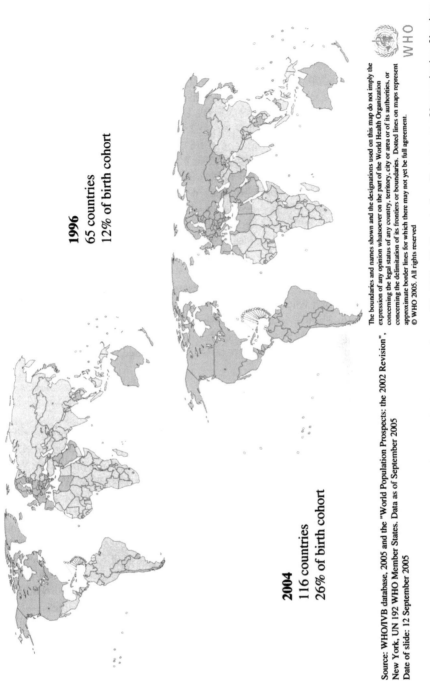

1996
65 countries
12% of birth cohort

2004
116 countries
26% of birth cohort

The boundaries and names shown and the designations used on this map do not imply the expression of any opinion whatsoever on the part of the World Health Organization concerning the legal status of any country, territory, city or area or of its authorities, or concerning the delimitation of its frontiers or boundaries. Dotted lines on maps represent approximate border lines for which there may not yet be full agreement.
© WHO 2005. All rights reserved

WHO

Source: WHO/IVB database, 2005 and the "World Population Prospects: the 2002 Revision", New York, UN 192 WHO Member States. Data as of September 2005
Date of slide: 12 September 2005

Fig. 1 Countries using rubella vaccine in their national immunization system. *Source:* World Health Organisation, Department of Immunisation, Vaccines and Biologicals. Reproduced with kind permission.

Incorporation of rubella vaccine in the national immunization program is also related to a country's economic status. In 1996, none of the 46 least developed countries have introduced rubella vaccine, compared with 50% (13/26) of countries with economies in transition (Eastern European countries and the newly independent states of the former Soviet Union), 60% (53/88) of developing countries, 74% (20/27) of territories, protectorates or self-governing areas and 93% (25/27) of developed countries.

Cost-effectiveness of rubella-containing vaccines in developing countries

Globally, rubella vaccination programs have not only resulted in significant reduction of morbidity and mortality, but cost savings. Recently, Hinman et al. reviewed published information and program documents on economic analyses of rubella and rubella containing vaccines globally (Hinman et al., 2002). Twenty-two articles were found, of which five cost analyses and five cost–benefit analyses (CBA) were performed in developing countries, and five CBA of rubella vaccine, five CBA of MMR vaccine and two cost-effectiveness analyses (CEA) were performed in developed countries.

The benefit–cost (B:C) ratio for a routine childhood programs in three developed countries was estimated to be 5.8–11.1. One country estimated the B:C ratio to be 3.2 for a two-dose program over 20 years. Of the five cost analyses in developing countries, all studies were conducted in the Americas. The annual cost to treat an infant with CRS ranged from $2,291 in Panama (1989) to $13, 482 in Jamaica, whereas in Guyana, the lifetime cost to treat an infant with CRS was $63,990.

Because of the measles elimination initiative in the Americas, many countries had simultaneously administered rubella-containing vaccine with the measles vaccine as part of campaigns and routine immunization programs. CBAs attempt to place a monetary value on the benefits of an intervention. Thus, a comparison of the amount spent on the intervention is divided by the cost of that intervention. For the entire English-speaking Caribbean to interrupt the transmission of rubella and prevent CRS, the cost–benefit ratio was 13.3:1. Two countries (Barbados, Guyana) estimated their own costs for interruption of transmission. For Barbados, the B:C ratio was 4.7:1. For Guyana, the B:C ratio was 38.8 with cost-effectiveness ratio of $1,633 per CRS case prevented.

From these studies, rubella vaccine programs globally have been cost saving. These studies are important in assisting governments to decide where to invest limited resources. When compared with other vaccine-preventable diseases such as Hib and Hepatitis B, rubella vaccination programs were more cost saving. These studies can assist governments to assess whether or not to implement or to enhance their rubella control activities.

Summary

With the introduction of rubella vaccine in 1969, significant reduction in morbidity and mortality has occurred, breaking the cycle of rubella epidemics and preventing catastrophes like the pandemic witnessed in 1962–1965. The vaccine has proven to be both safe and effective with persistence of effectiveness for over 30 years.

With the eradication of smallpox and near eradication of polio, the focus is now the elimination of measles. The significant reduction in measles by vaccination in many countries and regions has made the burden of rubella more obvious. As a result, over the last 10 years the use of rubella vaccine together with measles vaccine has increased, although many populous countries are still exceptions. Nevertheless, the future looks positive for elimination from the world of rubella and its devastating consequences in the form of CRS.

References

Balfour Jr HH, Groth KE, Edelman CK, Amren DP, Best JM, Banatvala JE. Rubella viraemia and antibody responses after rubella vaccination and reimmunisation. Lancet 1981; 1: 1078–1080.

Best JM. Rubella vaccines: past, present and future. Epidemiol Infect 1991; 107: 17–30.

Bott LM, Eisenberg DH. Congenital rubella after successful vaccination. Med J Aust 1982; 1: 514–515.

Braun C, Kampa D, Fressle R, et al. Congenital rubella syndrome despite repeated vaccination of the mother: a coincidence of vaccine failure with failure to vaccinate. Acta Paediatr 1994; 83: 674–677.

Buimovici-Klein E, Cooper LZ. Cell-mediated immune response in rubella infections. Rev Infect Dis 1985; 7(Suppl 1): S123–S128.

CDC. Measles, mumps and rubella—vaccine use and strategies for elimination of measles, rubella and congenital rubella syndrome and control of mumps: recommendations of the Advisory Committee on Immunization Practices (ACIP). MMWR 1998; 47(RR-8): 36–37.

CDC. Notice to readers: revised ACIP recommendations for avoiding pregnancy after receiving a rubella-containing vaccine. MMWR 2001; 50: 1117.

Cutts FT, Robertson SE, Diaz-Ortega JL, Samuel R. Control of rubella and congenital rubella syndrome (CRS) in developing countries, Part 1: Burden of disease from CRS. Bull World Health Organ 1997; 75: 55–68.

Das BD, Lakhani P, Kurtz JB, et al. Congenital rubella after previous maternal immunity. Arch Dis Child 1990; 65: 545–546.

Davies WJ, Larson HE, Simsarian JP, et al. A study of rubella immunity and resistance to infection. JAMA 1971; 215: 600–608.

Dilraj A, Cutts FT, Bennett JV, Fernandez de Castro J, Cohen B, Coovadia HM. Persistence of measles antibody two years after revaccination by aerosol or subcutaneous routes. Pediatr Infect Dis J 2000; 19(12): 1211–1213.

Eilard T, Strannegard O. Rubella reinfection in pregnancy followed by transmission to the fetus. J Infect Dis 1974; 129: 594–596.

Enders G, Calm A, Schaub J. Rubella embryopathy after previous maternal rubella vaccination. Infection 1984; 12: 96–98.

Enders G. Serologic test combinations for safe detection of rubella infections. Rev Infect Dis 1985; 7(Suppl 1): S113–S122.

Fogel A, Gerichter CB, Barena B, et al. Response to experimental challenge in persons immunized with different rubella vaccines. J Pediatr 1978; 92: 26–29.

Forleo-Neto E, Carvalho ES, Fuentes IC, Precivale MS, Forleo LH, Farhat CK. Seroconversion of a trivalent measles, mumps, and rubella vaccine in children aged 9 and 15 months. Vaccine 1997; 15: 1898–1901.

Forsgren M, Calstrom G, Strangert K. Congenital rubella after maternal reinfection. Scand J Infect Dis 1979; 11: 81–83.

Furukawa T, Miyata T, Konda K, et al. Rubella vaccination during an epidemic. JAMA 1970; 213: 987–990.

Gilbert J, Kudesia G. Fetal infection after maternal reinfection with rubella. BMJ 1989; 299: 12–17.

Greenberg MA, Birx DL. Safe administration of mumps–measles–rubella vaccine in egg-allergic children. J Pediatr 1988; 13: 504–506.

Gregg NM. Congenital cataract following German measles in the mother. Trans Ophthalmol Soc Aust 1941; 3: 35–46.

Hillary IB. Trials of intranasally administered rubella vaccine. J Hyg (Camb) 1971; 69: 547–552.

Hinman AR, Irons B, Lewis M, Kandola K. Economic analyses of rubella and rubella vaccines: a global review. Bull World Health Organ 2002; 80: 264–270.

Hofmann J, Kortung M, Pustowoit B, et al. Persistent fetal rubella vaccine virus infection following inadvertent vaccination during early pregnancy. J Med Virol 2000; 61: 155–158.

Honeyman MC, Forrest JA, Dorman DC. Cell-mediated immune response following natural rubella and vaccination. Clin Exp Immunol 1974; 17: 665–671.

Horstmann DM, Schluederberg A, Emmons JE, Evans BK, Randolph MF, Andiman WA. Persistence of vaccine-induced immune responses to rubella: comparison with natural infection. Rev Infect Dis 1985; 7(Suppl 1): S80–S85.

Howson CP, Howe CJ, Fineberg HV. Chronic Arthritis in Adverse Effects of Pertussis and Rubella Vaccines. Washington, DC: Institute of Medicine, National Academy; 1991; pp. 187–205.

Kakizawa J, Nitta Y, Yamashita T, Ushijima H, Katow S. Mutations of rubella virus vaccine TO-336 strain occurred in the attenuation process of wild progenitor virus. Vaccine 2001; 19: 2793–2802.

Keith CG. Congenital rubella infection from reinfection of previously immunized mothers. Aust N Z J Ophthalmol 1991; 19: 291–293.

Kelso JM, Jones RT, Yunginger JW. Anaphylaxis to measles, mumps, and rubella vaccine medicated by IgE to gelatin. J Allergy Infect Dis 1993; 91: 867–872.

Lalla M, Vesikari T, Virolainen M. Lymphoblast proliferation and humoral antibody response after rubella vaccination. Clin Exp Immunol 1973; 15: 192–202.

Landrigan PL, Stoffels MA, Anderson E, Witte JJ. Epidemic rubella in adolescent boys: clinical features and results of vaccination. JAMA 1974; 227: 1283–1288.

Lavi S, Zimmermann B, Koren G, Gold R. Administration of measles, mumps, and rubella virus vaccine (live) to egg-allergic children. JAMA 1990; 263: 269–271.

Meyer Jr HM, Parkman PD, Hobbins TE, Larsons E, Davis WJ, Simsarian JP, Hopps HE. Attenuated rubella virus: laboratory and clinical characteristics. Am J Dis Child 1969; 118: 155–163.

Mitchell L, Ho M, Rogers J, et al. Rubella reimmunization: comparative analysis of the immunoglobulin G response to rubella virus vaccine in previously seronegative and seropositive individuals. J Clin Microbiol 1996; 34: 2210–2218.

Mitchell LA, Tingle AJ, Shukin R, Sangeorzan JA, McCune J, Braun DK. Chronic rubella vaccine-associated arthropathy. Arch Intern Med 1993; 153: 2268–2274.

Morgan-Capner P, Miller E, Vurdien J, Ramsay M. Outcome of pregnancy after maternal reinfection with rubella. Commun Dis Rep 1991; 1: R57–R59.

O'Shea S, Corbett K, Barrow S, et al. Rubella reinfection: role of neutralizing antibodies and cell-mediated immunity. Clin Diagn Virol 1994; 2: 349–358.

Paludetto R, van den Huevel J, Stagni A, et al. Rubella embrypopathy after maternal infection. Biol Neonate 1994; 65: 340–341.

Parkman PD, Buescher EL, Artenstein MS. Recovery of rubella virus from army recruits. Proc Soc Exp Biol Med 1962; 111: 225–230.

Parkman PD, Meyer HM, Kirschstein BL, Hopps HE. Attenuated rubella virus. I. Development and laboratory characterization. N Engl J Med 1966; 275: 569–574.

Perkins FT. Licensed vaccines. Rev Infect Dis 1985; 7(Suppl 1): S73–S76.

Plotkin SA, Farquahr JD, Katz M, Buser F. Attenuation of RA 27/3 rubella virus in WI-38 human diploid cells. Am J Dis Child 1969; 118: 178–185.

Plotkin SA. Rubella vaccine. In: Vaccines (Plotkin SA, Orenstein WA, editors). 4th ed., Chapter 26. Philadelphia: W.B. Saunders Co.; 2004; pp. 703–743.

Pougatcheva SO, Abernathy ES, Vzorov AN, Compans RW, Frey TK. Development of a rubella virus DNA vaccine. Vaccine 1999; 17: 2104–2112.

Preblud SR, Serdula MK, Frank Jr JA, Brandling-Bennett AD, Hinman AR. Rubella vaccination in the United States: a ten-year review. Epidemiol Rev 1980; 2: 171–194.

Puschak R, Young M, McKee TV, Plotkin SA. Intranasal vaccination with RA 27/3 attenuated rubella virus. J Pediatr 1971; 79: 55–70.

Ray P, Black S, Shinefield H, et al. Risk of chronic arthropathy among women after rubella vaccination. JAMA 1997; 278: 551–556.

Rubella Surveillance. National Communicable Disease Center, United States Department of Health, Education and Welfare (No.1); June 1969.

Sakaguchi M, Ogura H, Inouye S. IgE antibody to gelatin in children with immediate-type reactions to measles and mumps vaccines. J Allergy Infect Dis 1995; 96: 563–565.

Scott HD, Byrne EB. Exposure of susceptible pregnant women to rubella vaccines: serological findings during the Rhode Island immunization campaign. JAMA 1971; 215: 609–612.

Sepulveda J, Valdespino JL, personal communication 2001.

Sever JL, Schiff GM, Huebner RJ. Inactivated rubella virus vaccine. J Lab Clin Med 1963; 62: 1015.

Singh R, John TJ, Cherian T, Raghupathy P. Immune response to measles, mumps & rubella vaccine at 9, 12 & 15 months of age. Indian J Med Res 1994; 100: 155–159.

Slater P, Ben-Zvi T, Fogel A, et al. Absence of an association between rubella vaccination and arthritis in underimmune post-partum women. Vaccine 1995; 13: 1529–1532.

Tingle A, Mitchell L, Grace M, et al. Randomised double-blind placebo-controlled study on adverse effects of rubella immunization in sero-negative women. Lancet 1996; 349: 1277–1281.

Tingle AJ, Allen M, Petty RE, Kettyls GD, Chantler JK. Rubella-associated arthritis. I. Comparative study of joint manifestations associated with natural rubella infection and RA 27/3 rubella immunization. Ann Rheum Dis 1986; 45: 110–114.

Tookey P. Rubella in England, Scotland and Wales. Euro Surveill 2004; 9: 21–23.

Veronelli JA. An open community trial of live rubella vaccines. Study of vaccine virus transmissibility and antigenic efficacy of three HPV-77 derivatives. JAMA 1970; 213: 1829–1836.

Waxman MN, Wolinsky JS. A model of the structural organization of rubella virions. Rev Infect Dis 1985; 7(Suppl 1): 133–139.

Weber B, Enders G, Schlosser R, et al. Congenital rubella syndrome after maternal reinfection. Infection 1993; 21: 118–121.

Weibel RE, Villarejos VM, Klein EB. Clinical laboratory studies of live attenuated RA 27/3 and HPV-77 DE rubella virus vaccine. Proc Soc Exp Biol Med 1980; 165: 44–49.

Weller TH, Neva FA. Propagation in tissue culture of cytopathic agents from patients with rubella-like illness. Proc Soc Exp Biol Med 1962; 111: 215–225.

WHO. Report of a Meeting on Preventing Congenital Rubella Syndrome: Immunization Strategies, Surveillance Needs. Geneva: WHO/V&B/00.10; 12–14 January 2000.

Yaru H, Zhao K, Yinxiang G, Sulan H, Shuzhen W, Changtai W. Rubella vaccine in the People's Republic of China. Rev Infect Dis 1985; 7(Suppl 1): S79.

Rubella Viruses
Jangu Banatvala and Catherine Peckham (Editors)
DOI 10.1016/S0168-7069(06)15005-3

Chapter 5

Rubella Epidemiology: Surveillance to Monitor and Evaluate Congenital Rubella Prevention Strategies

Catherine Peckham, Pat Tookey, Pia Hardelid

Centre for Paediatric Epidemiology and Biostatistics, UCL Institute of Child Health, 30 Guilford Street, London, WC1N 1EH, UK

Rubella infection usually presents as a mild or asymptomatic infection in children and adults and often goes undiagnosed. In pregnant women however, the consequences can be serious, especially when infection occurs in the first trimester. The virus is teratogenic and may cross the placenta causing fetal infection, which if acquired early in fetal life may be associated with serious consequences. The most commonly described congenital rubella anomalies include sensorineural hearing impairment, cataracts, cardiac defects, impaired fetal growth and mental retardation (see Chapter 2). The aim of vaccination programmes is therefore to prevent cases of congenital rubella. To effectively achieve this goal, comprehensive surveillance programmes to monitor immunity in the population and detect cases of both rubella and congenital rubella are needed along with an understanding of the likely outcomes of different vaccination strategies, which is not intuitive.

Congenital rubella infection

There is a paucity of reliable information on the prevalence of congenital rubella defects and disability, particularly in developing countries, due to the absence of routine congenital rubella surveillance and the difficulty of diagnosing congenital rubella as the cause of a defect after the first months of life. However, at least 100,000 or more cases occur each year worldwide (World Health Organisation, Department of Vaccines and Biologicals, 2000). Only a minority of affected infants present with the classic triad of clinical signs that are recognisable at birth, which is often referred to as congenital rubella syndrome (CRS). A retrospective diagnosis of congenital rubella is not possible when a child presents later in childhood with a defect compatible with congenital rubella, such as sensorineural hearing loss, as congenital infection cannot be distinguished from natural infection or vaccine-induced immunity. This is particularly so in a non-epidemic situation. Nevertheless, studies have shown that the incidence of congenital rubella defects in developing countries is at least as high as that reported in developed countries before the introduction of vaccine. Most estimates of the prevalence of congenital rubella defects following congenital infection are based on the outcome of pregnancies identified during a rubella epidemic. These studies may underestimate the risk of adverse effects as follow up is usually short term and isolated cases of sensorineural hearing loss are not likely to be identified. Based on a comprehensive review of the literature, the incidence of congenital rubella in developing countries during epidemics has been estimated to be 0.5–2.2 per 1000 live births (about 1.7 per 1000 live births in Israel, 0.7 in Oman, 2.2 in Panama, 1.5 in Singapore and 1.7 in Jamaica) (Cutts et al., 1997). These rates are comparable to those in industrialised countries in the pre-vaccine era. It is logical to assume therefore that the problem of congenital rubella is universal and that where rubella susceptibility in women of child-bearing age is high, rubella poses a serious problem.

In most countries data on the prevalence of congenital rubella defects in non-epidemic situations remains limited. Epidemiological studies carried out in the UK in the pre-vaccine era suggested that congenital rubella accounted for at least 16% of the cases of moderate to severe sensorineural deafness (Peckham et al., 1979) and about 2% of the cases of congenital heart disease among British children (Peckham, 1985). In France, it was estimated that 16% of congenital cataracts in children could be attributed to congenital rubella (Celers et al., 1983). Studies in developing countries, which have largely focused on specific conditions such as congenital cataracts or childhood deafness, have identified congenital rubella as an important potentially preventable cause of these defects (Eckstein et al., 1996). These studies highlight the need for establishing rubella vaccination programmes throughout the world.

Risk of congenital infection and risk of defects

When rubella infection is acquired in early pregnancy the risk of fetal infection with associated defects is high and a substantial number of infants are severely affected with multiple defects, particularly of the eye, heart and ear and neurological abnormalities (see Chapter 2). In contrast, fetal infection later in the third and fourth month usually results in sensorineural deafness alone. Most congenitally infected infants are affected as a result of primary maternal infection although damage resulting from asymptomatic or clinically apparent maternal reinfection in women

Table 1

Risk of infection, defect and overall risk of defects following serologically confirmed rubella at successive stages in pregnancy

Week	Risk of infection (%)	Risk of defect (%)	Overall risk (%)
2–10	100	89	89
11–12	73	50	37
13–16	52	33	17
17–18	38	7	3
19+	25–82	0	0

Source: Data from Miller et al. (1982, 1991).

with previous immunity has also been described (Miller, 1990; Morgan-Capner et al., 1991; Ushida et al., 2003). The risk of fetal infection following maternal reinfection during the first 12 weeks of pregnancy is thought to be low, probably less than 10% (Miller, 1990; Morgan-Capner et al., 1991); fetal infection is not likely to be associated with damage following subclinical maternal reinfection, but the risk might be higher following symptomatic reinfection, which is very rare.

In a large prospective study in the UK the pregnancy outcome of over 1000 women with laboratory confirmed rubella infection in pregnancy was established. However, over 90% of women infected in the first trimester of pregnancy and half of those with infection between 13 and 16 weeks of gestation chose to have a therapeutic abortion. The dates of the last menstrual period and rash were known, and the infants were followed up systematically for several years (Miller, 1991). Studies based on serological diagnosis of rubella in pregnancy with longer term follow-up are more likely to reflect the true risk of damage than those earlier studies based on a clinical diagnosis. Table 1 shows the risk of fetal infection by week of maternal infection, the risk of fetal damage in children with proven infection and the overall risk of damage in a pregnancy complicated by rubella.

Overall, around 80% of infants exposed to maternal rubella in the first 8 weeks of pregnancy had severe rubella-associated impairment including congenital heart defects, eye defects and hearing impairment.

Congenital rubella defects occurred in about 50% of infants exposed to maternal rubella in weeks 9–12 and at this stage hearing impairment was the most frequent defect. Exposure to maternal rubella at 13–16 weeks resulted in fewer than 30% of infants with damage and defects other than congenital hearing loss were unusual. Infections after 16 weeks of gestation were only occasionally associated with defects, although cases of deafness have been reported up to 22 weeks of gestation. Well-designed prospective studies of children born after later infection have shown that although rubella infection after this period can result in fetal infection, there appears to be no increased risk of defects (Miller et al., 1982; Grillner et al., 1983).

Preconception rubella

Although there have been isolated reports of possible congenital defects following maternal rubella before conception, in none of these cases was the maternal or fetal infection confirmed by adequate laboratory tests. In a prospective study, Enders et al. (1988) found no risk associated with exposure to rubella before conception.

Long-term manifestations of congenital rubella

It is difficult to establish the full impact of congenital rubella on a population, as there is not always a history of maternal infection and defects may not be apparent or even develop until weeks, months or years after birth, when a retrospective diagnosis of congenital infection is usually no longer possible (Peckham, 1972). Sensorineural hearing loss, the most frequent late-onset defect, may be moderate or severe, bilateral or unilateral and is often associated with pigmentary retinopathy. Other manifestations of late-onset disease include pneumonitis, diabetes mellitus, hypothyroidism, growth hormone deficiency and encephalitis (Marshall, 1972). Follow up of established cohorts of children with congenital rubella diagnosed at birth into adult life have demonstrated the progressive nature of the disease and its long-term sequelae. The follow-up in 1990 of children with congenital rubella born in the 1970s and 1980s and reported to the UK National Congenital Rubella Surveillance Programme demonstrated an increased incidence of diabetes and thyroid problems in adolescents with congenital rubella compared with the general population. This concurs with findings reported from the original Australian cohort where follow-up to age 60 years has been reported (Menser et al., 1967, 1978; Forrest et al., 1971; McIntosh and Menser, 1992; Forrest et al., 2002). There are reports of higher than expected rates of hearing loss, glaucoma, insulin-dependent diabetes, hypo- and hyperthyroidism, hormone imbalances, premature ageing and gastrointestinal and oesophageal problems, but these were mostly based on follow-up of small numbers of individuals with problems already manifested in early life. There is a paucity of information on the later development of children with congenital infection and no signs of rubella damage in early life. It has been suggested that an autoimmune process may play a role in some of these late-onset conditions (Cooper and Alford, 2001).

Epidemiological features of the rubella virus

The widespread use of rubella vaccine has had a marked impact on the rubella antibody profile of the population and the prevention of congenital rubella. In most countries, in the absence of rubella immunisation, rubella is endemic with epidemic cycles occurring every 4–7 years, usually in the spring or early summer in temperate climates, and the size of an outbreak depends on the number of susceptible individuals in the population and their clustering within the population. Seroepidemiological studies carried out in industrialised countries in the pre-vaccination era generally showed that the proportion of individuals with rubella antibodies increased steadily with age and that overall, about 80% of women of childbearing age had rubella antibodies and around 20% were still susceptible to infection. Most infections were acquired early in childhood, with around 50% of children seropositive by 6–8 years and a peak age of infection between 4 and 11 years. However, in some countries the rate of acquisition of antibodies was more rapid than in others, probably due to factors relating to crowding, social mixing and to different patterns of childcare. There were some notable exceptions where susceptibility rates remained high, including parts of Asia and Africa and some island populations (Miller, 1991). In these situations, the introduction of wild rubella virus into the community would result in rubella epidemics with serious consequences for pregnant women and their offspring.

Rubella susceptibility among pregnant women

The proportion of susceptible women, and hence the risk of rubella in pregnancy, may vary widely between population subsets within a country as well as between countries. For example, in a population-based seroprevalence study in the Netherlands blood samples were collected from over 8000 individuals, 20% of whom lived in municipalities with low uptake of vaccine. Serological profiles of these data, which were representative of the population, showed clusters of susceptible individuals within specific age, social and geographical groups (de Haas et al., 1999). A more recent study in the Netherlands demonstrated that although overall vaccination rates were high and herd immunity sufficient among the general population, this was not the case in orthodox reformed communities (de Melker et al., 2003). In 2004, a rubella outbreak within these communities led to several children being born with congenital rubella-related defects (van der Veen et al., 2005). In the UK, in the 1980s, a disproportionate number of congenitally infected infants were born to Asian women, a group known to have higher rates of rubella susceptibility than non-Asian women (Miller et al., 1991). In contrast, women most affected in the 1960s were recent immigrants from the Caribbean, where there had not been a rubella epidemic for many years and the prevalence of seropositivity was low.

More than 25 years after the introduction of rubella immunisation, analysis of blood samples collected routinely from women receiving antenatal care in North London showed that women from minority ethnic groups in the UK were more likely to be susceptible to rubella than white women (Table 2), a finding likely to reflect recent patterns of immigration and missed rubella vaccination in childhood; primiparous women were particularly likely to be susceptible (Tookey et al., 2002).

In these circumstances, if exposed to circulating rubella, women from minority ethnic groups are at increased risk of contracting rubella in pregnancy and having a child with congenital rubella. In the US, rubella now occurs mainly among foreign born Hispanic adults who are either unvaccinated or whose vaccination status is not known, and there is very little spread among the US resident population (Reef et al., 2002). These observations highlight the need for monitoring rubella

Table 2

Proportion of pregnant women susceptible to rubella by ethnic group and parity, North London 1996–1999

Ethnic group	Total number tested	Susceptible women (%)	Susceptible primiparous women (%)
White British	94,626	1.6	1.9
Other White	2480	3.4	3.8
Mediterranean	3715	3.7	4.4
Pakistani	4312	4.1	5.4
Indian	8469	4.4	6.7
Bangladeshi	2171	6.1	9.4
Sri Lankan	763	14.9	23.3
Other Asian	955	5.9	10.4
Oriental	1896	8.0	9.4
Black British	240	2.1	2.0
Black Caribbean	3242	3.1	3.2
Black African	4878	6.2	8.4
Other Black	651	4.3	6.5

Source: Data from Tookey et al. (2002).

seroprevalence within a country so that groups at risk of rubella infection in pregnancy, such as immigrants and refugees from countries with no immunisation programmes, can be identified and targeted immunisation programmes introduced.

Studies among pregnant women in developing countries have shown wide variation in the age-specific seroprevalence of rubella ranging from 2.7% to 68%. In an analysis of serological data from 45 developing countries the proportion of seronegative women was >25% in 12 countries, 10–24% in 20 countries and <10% in 13 countries (Cutts and Vynnycky, 1999).

Prevention of congenital rubella

Rubella vaccines

The last major epidemic in the US occurred in 1963–1964 and resulted in an estimated 13,000 fetal or neonatal deaths and 20,000 children born with congenital rubella-related defects (Cooper and Alford, 2001). This epidemic had enormous human and economic cost implications and was a major impetus for the development of a rubella vaccine and for the rapid implementation of vaccination programmes.

The primary purpose of rubella vaccination is to prevent congenital rubella infection (see Chapter 4). The first live-attenuated rubella vaccines were licensed for use in 1969 and with the introduction of rubella vaccine into national immunisation programmes congenital rubella became a preventable condition. Rubella vaccine is safe and highly effective with a vaccine efficacy of about 98%. The vaccine can be delivered in combination with measles and mumps vaccine as MMR vaccine, in the routine childhood immunisation programme.

Rubella vaccine in pregnancy

Rubella vaccine given during or shortly before conception carries a theoretical risk of congenital rubella infection. Rubella vaccine virus has been shown to cross the placenta and to infect the fetus, and vaccine virus has been isolated from products of conception. Rubella-specific IgM, which indicates congenital infection, is present in about 5% of infants born to rubella-susceptible mothers vaccinated around the time of conception. In the UK, 4 of 25 infants born to susceptible women, who were vaccinated more than 1 week after the last menstrual period and followed prospectively, had rubella IgM at birth, but none of 21 infants born following vaccination of their susceptible mothers up to 3 months before LMP. None of these infants, including the four with rubella IgM, had evidence of congenital rubella damage (Tookey, 2001). In a similar study in Germany, in 1 of 5 cases, vaccine virus was transmitted to the infant as evidenced by persistent fetal infection. In one case, virus shedding persisted for more than 8 months and sequence analysis carried out on virus isolates from amniotic fluid, cord blood and the infant's urine confirmed infection by the vaccine strain; the infant had no evidence of rubella defects (Hofmann et al., 2000). These results have been collated with findings from prospective studies in various countries and no infants born to seronegative women who received rubella vaccine within 3 months of conception have been identified with congenital rubella defects (Best and Banatvala, 2004). Based on data from these studies and using the upper 95% confidence limit the maximum theoretical risk of a woman receiving rubella vaccine up to 3 months before conception giving birth to a child with congenital rubella defects is <0.6% for women vaccinated with Cendehill or RA 27/3 rubella vaccine strains. When the focus was limited to women vaccinated with RA 27/3 rubella vaccine strain vaccine between 1 and 6 weeks after the last menstrual period, the maximum theoretical risk for susceptible women is <2.5%, but numbers are small and confidence limits wide. Follow up of infants exposed to rubella vaccination close to the time of conception has usually been short and has focused on birth outcome when only the more serious defects are likely to have been identified (Best and Banatvala, 2004). Nevertheless, based on a review of all available information, inadvertent rubella vaccination in pregnancy is no longer considered to be an indication for termination of pregnancy.

Vaccination strategies—theoretical and practical considerations

Several research groups have modelled the transmission dynamics of rubella virus to assess the impact of different vaccine strategies and levels of vaccination uptake on age at infection, population seroprevalence and the number of pregnancies affected by rubella. Many studies focus on the dichotomy between schoolgirl vaccination programmes, sometimes referred to as a selective vaccination strategy, and the vaccination of boys and girls in early life, i.e. a universal strategy (Knox, 1985). A selective vaccination strategy can never completely prevent all cases of congenital rubella infection, since virus transmission is maintained among young children. However, introducing a selective strategy will always serve to decrease the number of cases of congenital rubella infection. This is in contrast to the universal strategy, which interrupts transmission of infection among children. In the long term, high rates of vaccination coverage will result in a population with few susceptible individuals and an interruption of rubella virus circulation, thus protecting pregnant women from infection. However, if high vaccine levels are not maintained, then this can result in a reduced circulation of the virus thereby delaying the age at which infection occurs in susceptible (unvaccinated) individuals, which may lead to an increase in the number of cases of congenital rubella (Anderson and May, 1983; Knox, 1985; Anderson and Grenfell, 1986; Anderson and May, 1991). This is particularly the case in countries with an underlying low rate of transmission, which leaves a significant proportion of women susceptible in childbearing age. This shift in the age at risk was clearly demonstrated in recent rubella outbreaks in Brazil and Costa Rica (Castillo-Solorzano et al., 2003).

Assuming a vaccine effectiveness of over 95% and homogenous mixing, it is estimated that a vaccination coverage of at least 85% is required to achieve a critical level of immunity to prevent the spread of infection within the population (Anderson and May, 1983; Nokes and Anderson, 1988). However, even when high overall vaccine coverage is achieved, herd immunity can be compromised if there are groups of unvaccinated individuals within a community, clustered geographically on religious or socio-demographical grounds. Outbreaks of measles, mumps and rubella and clusters of congenital rubella cases have been documented in these situations (Briss et al., 1992; Mellinger et al., 1995; Hanratty et al., 2000; van der Veen et al., 2005).

This possible adverse effect of the introduction of a universal rubella vaccination programme needs to be taken into account when planning congenital rubella prevention strategies. In countries such as Ethiopia, where infection rates are high among young children, only a small proportion of women of childbearing age remain susceptible (Cutts et al., 2000). The introduction of rubella vaccine would not be appropriate unless very high levels of vaccine coverage could be maintained, which is unlikely in this and many other resource-poor settings, as it could result in an increase in the incidence of congenital rubella defects.

Despite recommendations made by the WHO to set up integrated national surveillance systems to monitor cases of rubella and congenital rubella, as well as the proportion of women of reproductive age who are seronegative (Cutts et al., 1999), few countries have an active congenital rubella surveillance system in place and given the inadequate uptake of MMR vaccination in many European countries' the eradication of congenital rubella is unlikely to be achieved in the near future. In some countries, such as Italy and Germany, MMR levels of vaccine uptake are not sufficient to interrupt rubella transmission in the community. In the UK, the recent decline in uptake of both doses of MMR vaccine (illustrated with data for England in Table 3) is also a cause for concern, particularly in London, where immunisation uptake for the 1st dose of MMR is only slightly higher than 50% in some localities (NHS Health and Social Care Information Centre, 2005).

Outbreaks of measles have already been reported in the UK (Jansen et al., 2003; Gupta et al., 2005) and across Europe. In Ireland, a measles outbreak in Dublin in 2000 led to 1115 reported cases, 355 hospital admissions and 3 deaths (McBrien et al., 2003). In early 2006, several measles outbreaks were reported in England, which led to hospitalisations and caused one death (Health Protection Agency, 2006). Several recent rubella epidemics have also been reported across Europe. In addition to the outbreak in the Netherlands mentioned above, an outbreak in Romania in 2002–2003 led to 115 000 reported cases (approximately 531/100 000 population) as well as 150 children born with suspected rubella-associated defects. Seven of these were laboratory confirmed (Rafila et al., 2004). These examples demonstrate that rubella virus is still circulating in Europe and can lead to outbreaks among unvaccinated groups. Although a single dose of vaccine with high coverage will produce prolonged periods of freedom from rubella, cohorts of susceptible individuals who did not receive vaccine or who failed to respond to a

Table 3

MMR uptake in England, by year of birthday

Year	1st dose by 2nd birthday (%)	1st dose by 5th birthday (%)	1st and 2nd dose by 5th birthday (%)
1994–1995	91		
1995–1996	92		
1996–1997	92		
1997–1998	91		
1998–1999	88		
1999–2000	88	93	76
2000–2001	87	92	75
2001–2002	84	91	74
2002–2003	82	90	75
2003–2004	80	90	75
2004–2005	81	89	73

Source: NHS Immunisation Statistics, England: 2004–2005 (NHS Health and Social Care Information Centre, 2005).

single dose, will accumulate over time, along with the risk of future epidemics, usually among older children and young adults. In view of this, all countries in the European region of the WHO now recommend two doses of MMR (Weekly Epidemiological Record, 2005).

There are few studies documenting waning immunity following vaccination or infection. However, with the diminished circulation of virus within the community and the lack of opportunity for immunological boosting of either vaccine or natural immunity through repeated exposure to infection, natural immunity may not be lifelong and vaccine-induced immunity may be lost even more rapidly than that resulting from natural infection (Davidkin et al., 2000). More women could then be susceptible to symptomatic reinfection and lower levels of maternal immunity could result in shorter duration of passive immunity in the infant. Seroprevalence studies are required to monitor age-specific rubella antibodies so that any decline in population immunity can be detected as early as possible. In a Swedish study, where a cohort of children aged 12 years was followed up for 8–16 years to monitor rubella immunity and seroconversions after the introduction of MMR, there was evidence of a loss of immunity with natural immunity persisting at higher levels than vaccine-induced immunity (Christenson and Bottiger, 1994).

The implementation of rubella vaccination programmes: global experiences

The first step in eliminating rubella infection is to build up a picture of the epidemiology of the infection through clinical and laboratory surveillance, and to determine the feasibility of achieving and maintaining high vaccine coverage to interrupt transmission while avoiding the risk of shifting the age of acquisition of rubella infection to older age groups. Targets for vaccination coverage need to be set. There also needs to be an assessment of the age- and sex-specific seroprevalence of rubella antibodies in the population, and where possible an estimate of the incidence of congenital rubella abnormalities. Inadequate use of rubella vaccine can do more harm than good; countries attempting rubella elimination should ensure that routine vaccine coverage in children is higher than 85% and sustainable in the long term, and that women of childbearing age are immune. Immunisation of children will alter the transmission dynamics and could lead to an increase in susceptibility in young women with the potential for an increase in the incidence of congenital rubella (Vynnycky et al., 2003).

Most rubella vaccination programmes were introduced following epidemics of rubella, which had highlighted the severity of the problem and heightened public and political awareness. In Poland, a rubella epidemic was reported in 1985–1986 in which many pregnant women were exposed and in 1988 rubella vaccine was introduced for 13-year-old schoolgirls (Zgorniak-Nowosielska et al., 1996). The initial approach, adopted by many countries, has been to immunise specific groups of individuals at risk of acquiring rubella in pregnancy, such as adolescent girls and rubella-susceptible women of childbearing age. The aim of this approach was to 'mop up' those who had escaped childhood infection rather than to interrupt transmission of infection within the population. It also allowed for the boosting of natural and vaccine-induced antibody by repeated exposure to infection thereby avoiding the potential risk of waning immunity in a vaccinated population. This approach had an immediate effect in reducing congenital rubella infection and in countries such as the UK, Australia and Canada was followed by the later introduction of universal immunisation for all children with MMR vaccine.

In the UK, rubella vaccination was introduced in 1970 for schoolgirls between 11 and 14 years of age. This was soon extended to include susceptible women of childbearing age, particularly those working with children. Pregnant women found to be susceptible to rubella at antenatal screening were offered vaccine postpartum. These policies resulted in a reduction in notifications of congenital rubella and rubella-associated terminations of pregnancy. Following the introduction in 1988 of MMR vaccine for both boys and girls in their second year of life, the 1994 mass campaign to immunise all school children aged 5–16 years with measles/rubella vaccine, and the addition of a second dose of MMR for pre-school children from 1996, circulation of wild rubella virus declined, and congenital rubella was virtually eliminated (Miller et al., 1997; Tookey and Peckham, 1999; Tookey 2004). The schoolgirl vaccination programme was discontinued in 1996, but routine antenatal testing for rubella antibody status with vaccination of susceptible women in the postpartum period is still recommended (HMSO, 1996).

In Greece, MMR vaccine was introduced in 1980 for boys and girls at 1 year with no policy relating to targets for vaccine (Panagiotopoulos et al., 1999). This almost certainly resulted in MMR being given mainly to children in the private sector with population coverage consistently below 50% and increased rates of susceptibility in older groups. In 1993, rubella incidence in young adults was higher than in any previous epidemic year. Twenty-five children were diagnosed with defects associated with congenital rubella infections, 24.6 per 100,000 live births, amounting to the largest epidemic since 1950. With the gradual increase in the proportion of susceptible women of childbearing age a subsequent outbreak occurred in the winter of 1998/1999 (Panagiotopoulos and Georgakopoulou, 2004), which resulted in around 2500 domestic cases (Giannakos et al., 2000). It also led to rubella being imported to the UK by Greek students attending British universities and led to the only congenital rubella birth reported in the UK in 1999 (Tookey et al., 2000).

Schoolgirl vaccination was introduced in Australia in 1971. This resulted in a significant decline in congenital rubella but cases continued to occur, even in those vaccinated, as a result of antibody loss or poor vaccine response. In view of this postpartum vaccination was introduced for women with low-level rubella antibodies detected at antenatal testing (Cheffins et al., 1998). In 1989, MMR vaccine was introduced for infants at 1 year and the adolescent programme for girls was discontinued in 1994–1995. In 1998 a second dose of MMR was recommended at 4 years. National surveillance systems were in place for the reporting of clinical and congenital rubella, which showed that in 1992–1995 the risk of rubella was > 20/100,000 and declined to 7.2 per 100,000 in 1997 (Sullivan et al., 1999).

In Finland, rubella immunisation for 11–13-year-old girls was introduced in 1975 with seroprevalence in women increasing to a plateau of 90% by 17 years of age. Following the introduction of MMR vaccination for young children in 1982 seroprevalence among children aged 2–3 years of age had increased to 95% by 1986. High vaccine coverage of over 97% has been sustained and rubella virtually eliminated with no cases of congenital rubella reported since 1986 (Davidkin et al., 2004).

In Sri Lanka (Gunasekera and Gunasekera, 1996), an epidemic of rubella occurred in 1994–1995 with 275 cases of congenital rubella reported in 1994 and 169 in 1995. Although rubella vaccine was then widely available in the private sector it was not available to the general population. Subsequently a national programme was introduced for all 11–15-year-old girls and, on request, for susceptible women aged 16–44 years.

In the USA, a different approach was taken and a policy of universal rubella immunisation was adopted in 1969. Rubella vaccine was given to boys and girls in their second year of life with the aim of developing herd immunity so that women who were susceptible to rubella would not be exposed, particularly to infection from their own children who posed the greatest risk. This approach had a dramatic effect and the wide use of rubella vaccine reduced the incidence of rubella and congenital rubella by >99% from pre-vaccine levels (Reef et al., 2000). In 1995, State Health Departments reported only 138 cases of rubella and 6 cases of CRS to CDC, demonstrating that the goal to eliminate indigenous congenital rubella by 2000 was an achievable target (Reef et al., 2002). In the US, rubella is nearing elimination but sporadic outbreaks, mainly among unvaccinated foreign residents, still pose challenges for its eradication (MMWR Morbidity Mortality Weekly Report, 2005).

Following the implementation of measles vaccination programmes in the Americas, regional measles surveillance systems documented widespread rubella virus circulating in many countries with a potential for major rubella epidemics (Castillo-Solorzano et al., 2003). As a result rubella prevention efforts were strengthened and the goal of rubella elimination added to the current measles immunisation programme. Countries in the Americas are at different stages of rubella prevention strategies but the majority now recommend the inclusion of MMR vaccine in their routine childhood programmes. In order to accelerate the congenital rubella prevention programme and reduce the number of women of childbearing age who are susceptible to rubella, many countries undertook one-off mass vaccination campaigns targeting females aged 5–39 years with measles and rubella-containing vaccines. In some countries both males and females were targeted to reduce the circulation of virus in the community and interrupt virus transmission. In order to assess the burden of rubella disease, integrated surveillance systems were developed for measles and rubella as well as for congenital rubella and laboratory reports of virus isolation (Castillo-Solorzano et al., 2003).

More recently in Brazil, and Costa Rica and the West Indies, mass immunisation programmes have been introduced to avoid outbreaks of measles and rubella. The impact of these combined rubella vaccination strategies and the rapid interruption of rubella virus transmission is already evident in Cuba, Brazil, Costa Rica, Chile and the English-speaking Caribbean (Massad et al., 1995; Castillo-Solorzano et al., 2003).

In 1999, there was a large rubella epidemic in Costa Rica with the highest attack rate in adults 20–39 years and around 500 cases occurring in pregnant women. This was likely to have resulted from the shift in age-specific incidence of infection resulting from the introduction of MMR vaccine in children. In response to this outbreak, women aged 18–40 were offered MMR vaccine in a campaign carried out in two phases: non-pregnant women and then pregnant women. Women were informed about the safety of the vaccine and pregnancy outcomes among women vaccinated in pregnancy were monitored.

A similar situation arose in Brazil where women of childbearing age were vaccinated in a mass campaign and those who were subsequently found to be pregnant are being systematically followed up and pregnancy outcome assessed (EPI Newsletter, 2002). From the information gathered in these investigations it should be possible to assess the safety of giving rubella vaccine to pregnant women. However, it will be essential to follow up the infants exposed to rubella vaccine *in utero* to determine their infection status and for those infected longer term follow-up will be required to identify defects that may not be apparent at birth.

In 1996 a survey on worldwide rubella vaccination use, sponsored by WHO, identified 78/214 (36%) countries with a national policy for rubella immunisation and inclusion of rubella vaccine in their national immunisation programmes. Seventy-eight countries had a national policy of rubella vaccination. This included 92% of industrialised countries and 12% of developing countries (Robertson et al., 1997). By 2004, this had increased to 116 (54%) of the 214 countries and territories reporting to the WHO (World Health Organisation, 2006).

Surveillance of rubella and congenital rubella

In order to monitor the efficacy of a rubella immunisation programme and to detect unforeseen problems that might arise, surveillance systems need to be in place. WHO developed guidelines for the surveillance of rubella and congenital rubella in 1999 (Cutts et al., 1999), which were updated in 2003 (World Health Organisation, 2003). These recommend clinical surveillance of rubella and congenital rubella infection; serological surveillance to detect changes in the prevalence of rubella immunity in the population; and the recording of vaccine coverage. The age- and sex-specific incidence of rubella infection and the incidence of congenital rubella infection should be made available at the country level.

As an example of national rubella surveillance, and to demonstrate the different components that need to be in place, we describe below the national surveillance of congenital rubella in the UK. Comparable systems are in operation in Canada, Australia, the US and other countries. Mass vaccination programmes change the epidemiological dynamics of rubella infection within the population and potential long-term effects, such as an increase in rubella susceptibility among women of childbearing age, should be anticipated by surveillance. Thus, a vaccination policy can be appropriately adapted to avoid complications.

Surveillance of clinical disease and seroprevalence

Laboratory confirmed cases of rubella are notified to the Health Protection Agency Centre for Infections. However, clinical and laboratory surveillance of rubella infection is of limited value as most infections are subclinical and will not be reported. Serological surveys are therefore used to monitor the prevalence of rubella antibodies in women of childbearing age in different sections of the community. These are usually based on antenatal sera, reflecting the seroprevalence of rubella-specific antibody in pregnant women, and provide information for policy decisions.

Surveillance of congenital rubella

The National Congenital Rubella Surveillance Programme (NCRSP) was established in 1971 to monitor the effect of the rubella vaccination policy in England, Scotland and Wales. Passive reporting through laboratory reporting systems and from audiologists and paediatricians was supplemented in 1990 by an active paediatric reporting scheme run through the British Paediatric Surveillance Unit of the Royal College of Paediatrics and Child Health (Nicoll et al., 2000). Data on maternal demographic characteristics and infection and vaccination history, as well as the clinical presentation in the infant and laboratory results, are routinely collected in order to monitor the risk factors associated with congenital rubella cases and to provide information on the changing epidemiology of congenital rubella in an era of prolonged high vaccination uptake.

The NCRSP data combined with routinely collected statistics from the Office for National Statistics on rubella-associated terminations of pregnancy demonstrate the dramatic decline in the number of pregnancies affected by rubella infection (see Fig. 1), from an average of about 40 births and 400 terminations reported each year in the period 1976–1980 to an average of about 4 births and 8 terminations a year in the 1990s. Data on rubella-associated terminations are no longer published since the annual number is so low (< 10). About a third of births reported since 1990 were to women who acquired their infection in early pregnancy outside the UK, and another third were to women who had arrived relatively recently in the UK and originated from countries without well-established immunisation programmes (Tookey, 2004).

This surveillance system has proved sensitive to small changes in incidence of congenital rubella. For example, in 1996 following a small outbreak of infection mainly among students, most cases were reported within 2 months of birth. Links with outbreaks elsewhere can also be observed, for example the one case reported in 1999 was associated with an outbreak in Greece (Tookey et al., 2000). However,

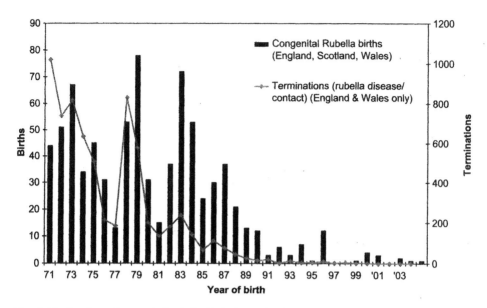

Fig. 1 Notifications of congenital rubella births in England, Scotland and Wales and notifications of rubella-associated terminations of pregnancy in England and Wales 1971–2005.

since the introduction of MMR, most of the cases reported have had multiple serious defects and it is likely that there has been an under ascertainment of single defects such as congenital rubella deafness. Deafness may not be detected until after the child receives MMR and the presence of vaccine immunity cannot easily be distinguished from the persistence of immunity resulting from congenital infection. Furthermore, as congenital rubella is so rare, it is unlikely to be considered as a potential explanation for isolated hearing loss, or other non-specific symptoms. As a result, in contrast to the period prior to the introduction of MMR vaccine, most reported cases are now severely symptomatic at birth. Between 1999 and 2005, 13 infants with congenital rubella were reported including one born in Ireland, and one stillborn infant. Nine were imported cases with maternal infection acquired abroad, and four were born to women who acquired the infection in the UK. Only two affected children had mothers who were born in the UK.

Conclusion

The introduction of rubella vaccine programmes has had a marked impact on the prevention of congenital rubella in developed countries, although cases still continue to occur. Few cases of rubella infection are now reported in the US or the UK, where surveillance systems are in place, and the few babies reported to have congenital infection are usually born to women from countries with no rubella vaccination programme or poor vaccine uptake rates (Reef et al., 2002; Tookey, 2004).

High rubella vaccine uptake must be maintained if rubella is to be eliminated. Furthermore, it is essential to have good surveillance systems in place so that changes in rubella susceptibility can be monitored and steps taken if susceptible groups start to emerge. There is also a need for greater public understanding of why vaccination programmes are important and the implications of low uptake.

The global burden of rubella and congenital rubella remains undefined in many developing countries. However, there is a risk that the incidence of rubella in women of childbearing age will increase in the absence of high vaccine coverage with a potential risk of adverse pregnancy outcomes and rubella-associated disability in surviving children. High levels of vaccine uptake need to be maintained if congenital rubella is to be eradicated. The European region of WHO have set a target for rubella elimination in the 52 countries of the region by 2010 (Weekly Epidemiological Record, 2005). There is still some way to go but until rubella is eradicated, unvaccinated individuals or communities will remain susceptible to imported infection, which could promote indigenous transmission. Even in countries with high vaccine coverage there is a risk of reintroduction of infection from neighbouring countries where vaccine uptake is not maintained. This is particularly pertinent for Europe where there is frequent travel across country borders. The heightened risk of rubella in immigrants from countries where vaccination is not universal has been highlighted in surveillance systems and reinforces the need for vaccination programmes to take this into account. Virus genotyping to identify the geographical origin of the infection is also a useful resource in surveillance programmes, particularly in countries such as the US and the UK where most cases of congenital rubella are due to imported infections.

More research is also needed regarding waning immunity following vaccination, particularly for the large cohorts of individuals in developed countries who have received their rubella vaccine at a very young age as part of an MMR immunisation programme. These countries have little circulating wild virus, and therefore no natural triggers for prolonged immunity. It is possible that rubella vaccine boosters are needed throughout life to prevent outbreaks of rubella in pregnant women in countries that have opted for the universal model of rubella vaccination.

References

Anderson RM, Grenfell BT. Quantitative investigations of different vaccination policies for the control of congenital rubella syndrome (CRS) in the United Kingdom. J Hyg (Lond) 1986; 96: 305–333.

Anderson RM, May RM. Vaccination against rubella and measles: quantitative investigations of different policies. J Hyg (Lond) 1983; 90: 259–325.

Anderson RM, May RM. Infectious Diseases of Humans: Dynamics and Control. Oxford: Oxford University Press; 1991.

Best J, Banatvala JE. Rubella. In: Principles and Practice of Clinical Virology (Zuckerman AJ, Banatvala JE, Pattison JR, Griffiths PD, Schoub BD, editors). Chichester: Wiley; 2004.

Briss PA, Fehrs LJ, Hutcheson RH, Schaffner W. Rubella among the Amish: resurgent disease in a highly susceptible community. Pediatr Infect Dis J 1992; 11: 955–959.

Castillo-Solorzano C, Carrasco P, Tambini G, Reef S, Brana M, de Quadros CA. New horizons in the control of rubella and prevention of congenital rubella syndrome in the Americas. J Infect Dis 2003; 187(Suppl 1): S146–S152.

Celers J, Aymard M, Feingold J, Godde-Jolly D. [Rubella and congenital cataracts]. Arch Fr Pediatr 1983; 40: 391–395.

Cheffins T, Chan A, Keane RJ, Haan EA, Hall R. The impact of rubella immunisation on the incidence of rubella, congenital rubella syndrome and rubella-related terminations of pregnancy in South Australia. Br J Obstet Gynaecol 1998; 105: 998–1004.

Christenson B, Bottiger M. Long-term follow-up study of rubella antibodies in naturally immune and vaccinated young adults. Vaccine 1994; 12: 41–45.

Cooper LZ, Alford CA. Rubella. In: Infectious Diseases of the Fetus and the Newborn Infant (Remington JS, Klein JO, editors). Philadelphia: W.B. Saunders Company; 2001; pp. 347–388.

Cutts FT, Abebe A, Messele T, Dejene A, Enquselassie F, Nigatu W, et al. Sero-epidemiology of rubella in the urban population of Addis Ababa, Ethiopia. Epidemiol Infect 2000; 124: 467–479.

Cutts FT, Best J, Siqueira MM, Engstrom K, Robertson SE. Guidelines for Surveillance of Congenital Rubella Syndrome (CRS) and Rubella. Geneva: World Health Organisation, Department of Vaccines and Other Biologicals; 1999.

Cutts FT, Robertson SE, Diaz-Ortega JL, Samuel R. Control of rubella and congenital rubella syndrome (CRS) in developing countries, Part 1: burden of disease from CRS. Bull World Health Organ 1997; 75: 55–68.

Cutts FT, Vynnycky E. Modelling the incidence of congenital rubella syndrome in developing countries. Int J Epidemiol 1999; 28: 1176–1184.

Davidkin I, Peltola H, Leinikki P. Epidemiology of rubella in Finland. Euro Surveill 2004; 9: 13–14.

Davidkin I, Peltola H, Leinikki P, Valle M. Duration of rubella immunity induced by two-dose measles, mumps and rubella (MMR) vaccination. A 15-year follow-up in Finland. Vaccine 2000; 18: 3106–3112.

de Haas R, van den HS, Berbers GA, de Melker HE, Conyn-van Spaendonck MA. Prevalence of antibodies against rubella virus in the Netherlands 9 years after changing from selective to mass vaccination. Epidemiol Infect 1999; 123: 263–270.

de Melker HE, van den HS, Berbers GA, Conyn-van Spaendonck MA. Evaluation of the national immunisation programme in the Netherlands: immunity to diphtheria, tetanus, poliomyelitis, measles, mumps, rubella and *Haemophilus influenzae* type b. Vaccine 2003; 21: 716–720.

Eckstein MB, Brown DW, Foster A, Richards AF, Gilbert CE, Vijayalakshmi P. Congenital rubella in south India: diagnosis using saliva from infants with cataract. BMJ 1996; 312: 161.

Enders G, Nickerl-Pacher U, Miller E, Cradock-Watson JE. Outcome of confirmed periconceptional maternal rubella. Lancet 1988; 1: 1445–1447.

EPI Newsletter. Brazil accelerates control of rubella and prevention of congenital rubella syndrome. EPI Newsl 2002; 24: 1–3.

Forrest JM, Menser MA, Burgess JA. High frequency of diabetes mellitus in young adults with congenital rubella. Lancet 1971; 2: 332–334.

Forrest JM, Turnbull FM, Sholler GF, Hawker RE, Martin FJ, Doran TT, et al. Gregg's congenital rubella patients 60 years later. Med J Aust 2002; 177: 664–667.

Giannakos G, Pirounaki M, Hadjichristodoulou C. Incidence of congenital rubella in Greece has decreased. BMJ 2000; 320: 1408.

Grillner L, Forsgren M, Barr B, Bottiger M, Danielsson L, De Verdier C. Outcome of rubella during pregnancy with special reference to the 17th–24th weeks of gestation. Scand J Infect Dis 1983; 15: 321–325.

Gunasekera DP, Gunasekera PC. Rubella immunisation-learning from developed countries. Lancet 1996; 347: 1694–1695.

Gupta RK, Best J, MacMahon E. Mumps and the UK epidemic 2005. BMJ 2005; 330: 1132–1135.

Hanratty B, Holt T, Duffell E, Patterson W, Ramsay M, White JM, et al. UK measles outbreak in non-immune anthroposophic communities: the implications for the elimination of measles from Europe. Epidemiol Infect 2000; 125: 377–383.

Health Protection Agency. Increase in measles cases in 2006, in England and Wales. CDR Wkly 2006; 16: 3.

HMSO. Immunisation against Infectious Disease. London: HMSO, Department of Health; 1996.

Hofmann J, Kortung M, Pustowoit B, Faber R, Piskazeck U, Liebert UG. Persistent fetal rubella vaccine virus infection following inadvertent vaccination during early pregnancy. J Med Virol 2000; 61: 155–158.

Jansen VA, Stollenwerk N, Jensen HJ, Ramsay ME, Edmunds WJ, Rhodes CJ. Measles outbreaks in a population with declining vaccine uptake. Science 2003; 301: 804.

Knox EG. Theoretical aspects of rubella vaccination strategies. Rev Infect Dis 1985; 7(Suppl 1): S194–S197.

Marshall WC. The clinical impact of intrauterine rubella. Ciba Found Symp 1972; 10: 3–22.

Massad E, Azevedo-Neto RS, Burattini MN, Zanetta DM, Coutinho FA, Yang HM, et al. Assessing the efficacy of a mixed vaccination strategy against rubella in Sao Paulo, Brazil. Int J Epidemiol 1995; 24: 842–850.

McBrien J, Murphy J, Gill D, Cronin M, O'Donovan C, Cafferkey MT. Measles outbreak in Dublin, 2000. Pediatr Infect Dis J 2003; 22: 580–584.

McIntosh ED, Menser MA. A fifty-year follow-up of congenital rubella. Lancet 1992; 340: 414–415.

Mellinger AK, Cragan JD, Atkinson WL, Williams WW, Kleger B, Kimber RG, et al. High incidence of congenital rubella syndrome after a rubella outbreak. Pediatr Infect Dis J 1995; 14: 573–578.

Menser MA, Dods L, Harley JD. A twenty-five-year follow-up of congenital rubella. Lancet 1967; 2: 1347–1350.

Menser MA, Forrest JM, Bransby RD. Rubella infection and diabetes mellitus. Lancet 1978; 1: 57–60.

Miller CL. Rubella in the developing world. Epidemiol Infect 1991; 107: 63–68.

Miller E. Rubella reinfection. Arch. Dis Child 1990; 65: 820–821.

Miller E. Rubella in the United Kingdom. Epidemiol Infect 1991; 107: 31–42.

Miller E, Cradock-Watson JE, Pollock TM. Consequences of confirmed maternal rubella at successive stages of pregnancy. Lancet 1982; 2: 781–784.

Miller E, Waight P, Gay N, Ramsay M, Vurdien J, Morgan-Capner P, et al. The epidemiology of rubella in England and Wales before and after the 1994 measles and rubella vaccination campaign: fourth joint report from the PHLS and the National Congenital Rubella Surveillance Programme. Commun Dis Rep CDR Rev 1997; 7: R26–R32.

Miller E, Waight PA, Vurdien JE, White JM, Jones G, Miller BH, et al. Rubella surveillance to December 1990: a joint report from the PHLS and National Congenital Rubella Surveillance Programme. CDR (Lond Engl Rev) 1991; 1: R33–R37.

MMWR Morbidity Mortality Weekly Report. Elimination of rubella and congenital rubella syndrome—United States, 1969–2004. MMWR Morb Mortal Wkly Rep 2005; 54: 279–282.

Morgan-Capner P, Miller E, Vurdien JE, Ramsay ME. Outcome of pregnancy after maternal reinfection with rubella. CDR (Lond Engl Rev) 1991; 1: R57–R59.

NHS Health and Social Care Information Centre. NHS Immunisation Statistics, England: 2004–05, Department of Health; 2005.

Nicoll A, Lynn R, Rahi J, Verity C, Haines L. Public health outputs from the British Paediatric Surveillance Unit and similar clinician-based systems. J R Soc Med 2000; 93: 580–585.

Nokes DJ, Anderson RM. Measles, mumps, and rubella vaccine: what coverage to block transmission? Lancet 1988; 2: 1374.

Panagiotopoulos T, Antoniadou I, Valassi-Adam E. Increase in congenital rubella occurrence after immunisation in Greece: retrospective survey and systematic review. BMJ 1999; 319: 1462–1467.

Panagiotopoulos T, Georgakopoulou T. Epidemiology of rubella and congenital rubella syndrome in Greece, 1994–2003. Euro Surveill 2004; 9: 17–19.

Peckham C. Congenital rubella in the United Kingdom before 1970: the prevaccine era. Rev Infect Dis 1985; 7(Suppl 1): S11–S16.

Peckham CS. Clinical and laboratory study of children exposed *in utero* to maternal rubella. Arch Dis Child 1972; 47: 571–577.

Peckham CS, Martin JA, Marshall WC, Dudgeon JA. Congenital rubella deafness: a preventable disease. Lancet 1979; 1: 258–261.

Rafila A, Marin M, Pistol A, Nicolaiciuc D, Lupulescu E, Uzicanin A, et al. A large rubella outbreak, Romania—2003. Euro Surveill 2004; 9: 7–9.

Reef SE, Frey TK, Theall K, Abernathy E, Burnett CL, Icenogle J, et al. The changing epidemiology of rubella in the 1990s: on the verge of elimination and new challenges for control and prevention. JAMA 2002; 287: 464–472.

Reef SE, Plotkin S, Cordero JF, Katz M, Cooper L, Schwartz B, et al. Preparing for elimination of congenital Rubella syndrome (CRS): summary of a workshop on CRS elimination in the United States. Clin Infect Dis 2000; 31: 85–95.

Robertson SE, Cutts FT, Samuel R, Diaz-Ortega JL. Control of rubella and congenital rubella syndrome (CRS) in developing countries, Part 2: vaccination against rubella. Bull World Health Organ 1997; 75: 69–80.

Sullivan EM, Burgess MA, Forrest JM. The epidemiology of rubella and congenital rubella in Australia, 1992 to 1997. Commun Dis Intell 1999; 23: 209–214.

Tookey P. Pregnancy is contraindication for rubella vaccination still. BMJ 2001; 322: 1489.

Tookey P. Rubella in England, Scotland and Wales. Euro Surveill 2004; 9: 21–22.

Tookey P, Molyneaux P, Helms P. UK case of congenital rubella can be linked to Greek cases. BMJ 2000; 321: 766–767.

Tookey PA, Cortina-Borja M, Peckham CS. Rubella susceptibility among pregnant women in North London, 1996–1999. J Public Health Med 2002; 24: 211–216.

Tookey PA, Peckham CS. Surveillance of congenital rubella in Great Britain, 1971–96. BMJ 1999; 318: 769–770.

Ushida M, Katow S, Furukawa S. Congenital rubella syndrome due to infection after maternal antibody conversion with vaccine. Jpn J Infect Dis 2003; 56: 68–69.

van der Veen Y, Hahne S, Ruijs H, van Binnendijk R, Timen A, van Loon A, et al. Rubella outbreak in an unvaccinated religious community in the Netherlands leads to cases of congenital rubella syndrome. Euro Surveill 2005; 10: 274–275.

Vynnycky E, Gay NJ, Cutts FT. The predicted impact of private sector MMR vaccination on the burden of congenital rubella syndrome. Vaccine 2003; 21: 2708–2719.

Weekly Epidemiological Record. Progress towards elimination of measles and prevention of congenital rubella infection in the WHO European Region, 1990–2004. Wkly Epidemiol Rec 2005; 80: 66–71.

World Health Organisation, Department of Vaccines and Biologicals. Report of a Meeting on Preventing Congenital Rubella Syndrome: Immunization Strategies, Surveillance Needs. Geneva: World Health Organisation; 2000.

World Health Organisation. Surveillance Guidelines for Measles and Congenital Rubella Infection in the WHO European Region. Copenhagen: WHO, Europe; 2003.

World Health Organisation. Rubella and Congenital Rubella Syndrome. http://www.who.int/immunization_monitoring/diseases/rubella/en/. Accessed 8th May 2006. (Electronic Citation).

Zgorniak-Nowosielska I, Zawilinska B, Szostek S. Rubella infection during pregnancy in the 1985–86 epidemic: follow-up after seven years. Eur J Epidemiol 1996; 12: 303–308.

Future Requirements

Successful rubella vaccination programmes have resulted in the virtual elimination of postnatally and congenitally acquired rubella in many countries (see Chapter 4 & 5). However, countries wishing to eliminate rubella must not only ensure high vaccine uptake levels, but programmes should be supported by high-quality surveillance, including molecular epidemiological studies, in order to obtain evidence about the circulation of indigenous viruses, or the importation of new strains from other parts of the world. This is particularly relevant when monitoring the later stages of rubella elimination.

Surveillance is dependent on not only employing techniques appropriate to local health services, but clinical and laboratory investigations must be monitored independently by those with appropriate experience.

There are still a number of countries, particularly in Sub Saharan Africa, in which there are no programmes for rubella elimination. However, it is important that such countries carry out epidemiological studies in order to determine the prevalence of rubella antibodies in different populations, and assess, if possible, the impact of congenitally acquired rubella on health services.

The development of robust laboratory investigations, which can be used in the field at ambient temperatures in developing countries, should be a priority, particularly for rubella-specific IgG and IgM, e.g. as has been accomplished for Dengue.

As infectious diseases become increasingly rare as a result of successful immunisation programmes, there is a tendency for more attention to be focused by the media on rare or even unrelated vaccine side effects. Unfortunately, unfounded and sensationalist reporting by the media may discourage vaccine uptake, as has been the case with MMR in the UK. This has resulted in a marked decline in vaccine uptake rates in recent years. WHO's Vaccination Safety Advisory Committee and national bodies, e.g. in the UK the Department of Health, The Medical Research Council and The Health Protection Agency, must be prepared to provide reliable, independent and scientific assessment on vaccine safety issues, preparing reports and liaising with the media within as short a time as possible.

Some of those involved in vaccination programmes continue to express concern about long-term protection among those with low or no longer detectable serum antibody levels some years after vaccination, particularly in countries where exposure to naturally acquired rubella becomes increasingly uncommon. Although challenge studies (see chapter 4) suggest that such persons are immune, research directed towards determining B- and T-cell immunity would be of value since this may result in eliminating the unnecessary expense of giving booster doses of vaccine to those with low or no longer detectable antibody responses.

Although current rubella vaccines are safe, well tolerated and immunogenic, this should not be a deterrent from developing new vaccines; aerosolised preparations show promise and obviate the need to use needles with the attendant risks of transmitting blood-borne infections in developing countries (see chapter 4).

List of Contributors

Jangu E. Banatvala
King's College London Medical
School at Guy's, King's College and
St Thomas' Hospitals
London, SEI 7EH, UK

Jennifer M. Best
King's College London School of
Medicine at Guy's, King's College and
St Thomas' Hospitals
London, SEI 7EH, UK

Min-Hsin Chen
Division of Viral Diseases
Centers for Disease Control and Prevention
1600 Clifton Road
Atlanta, GA 30333, USA

Gisela Enders
Institute of Virology
Infectiology and Epidemiology e.V and Labor
Prof. Dr. med Gisela Enders and Partner
Rosenbergstraße 85
70193, Stuttgart, Germany

Pia Hardelid
Centre for Paediatric Epidemiology and Biostatistics
UCL Institute of Child Health
30 Guilford Street
London, WCIN 1EH, UK

Joseph Icenogle
Division of Viral Diseases
Centers for Disease Control and Prevention
1600 Clifton Road
Atlanta, GA 30333, USA

Catherine Peckham
Centre for Paediatric Epidemiology and Biostatistics
UCL Institute of Child Health
30 Guilford Street
London, WCIN 1EH, UK

Stanley A. Plotkin
University of Pennsylvania and Sanofi Pasteur
4650 Wismer Road
Doylestown, PA 18901, USA

Susan Reef
Rubella and Mumps Activity
National Immunization Program
Centers for Disease Control and Prevention
1600 Clifton Road
Atlanta, GA 30333, USA

Pat Tookey
Centre for Paediatric Epidemiology and Biostatistics
UCL Institute of Child Health
30 Guilford Street
London, WCIN 1EH, UK

Index

Page numbers suffixed by t and f refer to Tables and Figures respectively.

Printed and bound by CPI Group (UK) Ltd, Croydon, CR0 4YY

03/10/2024

01040419-0020